27th Joint Seminar Development of Materials Science in Research and Education 2017

AA006699

IOP Conference Series: Materials Science and Engineering
Volume 266

Kezmarske Zlaby, Slovakia
4 – 8 September 2017

ISBN: 978-1-5108-5265-5
ISSN: 1757-8981

Printed from e-media with permission by:

Curran Associates, Inc.
57 Morehouse Lane
Red Hook, NY 12571

Some format issues inherent in the e-media version may also appear in this print version.

Copyright© (2017) by the Institute of Physics
All rights reserved. The material featured in this book is subject to
IOP copyright protection, unless otherwise indicated.

Printed by Curran Associates, Inc. (2017)

For permission requests, please contact the Institute of Physics
at the address below.

Institute of Physics
Dirac House, Temple Back
Bristol BS1 6BE UK

Phone: 44 1 17 929 7481
Fax: 44 1 17 920 0979

techtracking@iop.org

Additional copies of this publication are available from:

Curran Associates, Inc.
57 Morehouse Lane
Red Hook, NY 12571 USA
Phone: 845-758-0400
Fax: 845-758-2633
Email: curran@proceedings.com
Web: www.proceedings.com

Table of contents

Volume 266

27th Joint Seminar Development of Materials Science in Research and Education

4–8 September 2017, Kežmarské Žlaby, High Tatras, Slovakia

Accepted papers received: 6 November 2017
Published online: 23 November 2017

Preface

Papers

IOP Conf. Series: Materials Science and Engineering **266** (2017) 011001 doi:10.1088/1757-899X/266/1/011001

Preface

The seminar "Development of Materials Science in Research and Education", held on 4-8 September 2017 in Kežmarské Žlaby (Slovakia-High Tatras), was already the twenty seventh in the series started at Gabčíkovo in 1991 by the initiative of the Czech and Slovak Association for Crystal Growth and the Slovak Expert Group of Solid State Chemistry and Physics. The objective of this meeting offered an opportunity to Czech and Slovak teachers and scientists as well as guests from other countries who are working in the field of Materials Science to present their recent results and experience and to exchange new ideas and information.

The scientific sessions covered the following topics on materials science:

- Trends in development of materials research
- Education of materials science at the universities
- Information about the research programmes of individual institutions
- Information about equipment for preparation and characterization of materials
- Results of materials research

This workshop was aimed at creation of a stimulating atmosphere of cooperation and at the support of patient dissemination of scientific ideas and propagation of materials science in education. At the seminar, 38 participants presented 36 interesting talks.

Editors

Content from this work may be used under the terms of the Creative Commons Attribution 3.0 licence. Any further distribution of this work must maintain attribution to the author(s) and the title of the work, journal citation and DOI.
Published under licence by IOP Publishing Ltd

IOP Conf. Series: Materials Science and Engineering **266** (2017) 011002 doi:10.1088/1757-899X/266/1/011002

27th Joint Seminar

Development of Materials Science in Research and Education

Kežmarské Žľaby, 4 - 8 September 2017

Organized by

Slovak Expert Group of Solid State Chemistry and Physics
Czechoslovak Association for Crystal Growth

Under the auspices of

Crystallographic Society
Faculty of Chemical and Food Technology STU
Slovak Society for Industrial Chemistry
Faculty of Materials Science and Technology in Trnava STU
Regional Committee of Czech and Slovak Crystallographers
Institute of Physics, the Czech Academy of Sciences

Content from this work may be used under the terms of the Creative Commons Attribution 3.0 licence. Any further distribution of this work must maintain attribution to the author(s) and the title of the work, journal citation and DOI.
Published under licence by IOP Publishing Ltd

IOP Conf. Series: Materials Science and Engineering **266** (2017) 011002 doi:10.1088/1757-899X/266/1/011002

Scientific Committee

Marian Koman – *chairman*	Slovak University of Technology in Bratislava, Slovakia
Mária Behúlová	Slovak University of Technology in Bratislava, MTF Trnava, Slovakia
Zdeněk Kožíšek	Institute of Physics CAS, Prague, Czech Republic
Blažena Papánková	Slovak University of Technology in Bratislava, Slovakia

Organizing Committee

Ľubor Dlháň	Slovak University of Technology in Bratislava, Bratislava
Vladimír Jorík	Slovak University of Technology in Bratislava, Bratislava
Marian Koman	Slovak University of Technology in Bratislava, Bratislava
Blažena Papánková	Slovak University of Technology in Bratislava, Bratislava

SPONSORS OF DMSRE 2017 SEMINAR

Slovak Society for Industrial Chemistry, Koľová 15, 815 94 Bratislava, Slovak Republic

The Conference is organized with the support of the Ministry of Education, Science, Research and Sport of the Slovak Republic within the Research and Development Operational Programme for the project "University Science Park of STU Bratislava", ITMS 26240220084, co-funded by the European Regional Development Fund.

IOP Conf. Series: Materials Science and Engineering **266** (2017) 011003 doi:10.1088/1757-899X/266/1/011003

Peer review statement

All papers published in this volume of *IOP Conference Series: Materials Science and Engineering* have been peer reviewed through processes administered by the proceedings Editors. Reviews were conducted by expert referees to the professional and scientific standards expected of a proceedings journal published by IOP Publishing.

Content from this work may be used under the terms of the Creative Commons Attribution 3.0 licence. Any further distribution of this work must maintain attribution to the author(s) and the title of the work, journal citation and DOI.
Published under licence by IOP Publishing Ltd

IOP Conf. Series: Materials Science and Engineering **266** (2017) 012001 doi:10.1088/1757-899X/266/1/012001

Nucleation and microstructure development in Cr-Mo-V tool steel during gas atomization

M Behúlová, P Grgač and R Čička

Slovak University of Technology in Bratislava, Faculty of Materials Science and Technology in Trnava, Ulica Jána Bottu č. 2781/25, 917 24 Trnava, Slovakia

maria.behulova@stuba.sk

Abstract. Nucleation studies of undercooled metallic melts are of essential interest for the understanding of phase selection, growth kinetics and microstructure development during their rapid non-equilibrium solidification. The paper deals with the modelling of nucleation processes and microstructure development in the hypoeutectic tool steel Ch12MF4 with the chemical composition of 2.37% C, 12.06 % Cr, 1.2% Mo, 4.0% V and balance Fe [wt. %] in the process of nitrogen gas atomization. Based on the classical theory of homogeneous nucleation, the nucleation temperature of molten rapidly cooled spherical particles from this alloy with diameter from 40 μm to 600 μm in the gas atomization process is calculated using various estimations of parameters influencing the nucleation process – the Gibbs free energy difference between solid and liquid phases and the solid/liquid interfacial energy. Results of numerical calculations are compared with experimentally measured nucleation temperatures during levitation experiments and microstructures developed in rapidly solidified powder particles from the investigated alloy.

1. Introduction

Nucleation studies in the undercooled metallic melts are of essential interest for the understanding of phase selection and microstructure development. In addition to the knowledge of growth kinetics, the understanding of nucleation phenomena is an important ingredient for the interpretation of metastable phases formation because of its significant influence on the initial stage of solidification [1-6].

During the past decades, the positive consequences of rapid solidification of melts have been successfully applied in several progressive technologies of production and treatment of materials, e. g. solidification on cooling substrates, gas and water atomization techniques, plasma spray deposition, local surface laser or electron beam melting, etc.

High cooling rates of the melt enable to attain large melt undercoolings below the equilibrium liquidus temperature [7-9]. Consequently, deep initial undercoolings initiate high velocities of solid-liquid (S-L) interface and create conditions for the development of a great variety of materials with chemical compositions, solidification microstructures and properties which are impossible to produce by conventional metallurgical technologies. Generally, the increase in initial melt undercooling leads to the enhancement of homogeneity of solid phase, microstructure refinement, to the formation of metastable crystalline and quasi-crystalline phases, and also metallic glasses [1,10-11].

Introduced benefits of rapid solidification processing result from transition from equilibrium or quasi-equilibrium solidification in bulk volumes of melt to the rapid non-equilibrium solidification in spatially limited ranges by high S-L interfacial velocities. The development of final solidification

Content from this work may be used under the terms of the Creative Commons Attribution 3.0 licence. Any further distribution of this work must maintain attribution to the author(s) and the title of the work, journal citation and DOI.
Published under licence by IOP Publishing Ltd

microstructures is determined particularly by thermodynamic and thermokinetic conditions of rapid cooling, nucleation and solidification of melt [12-13].

Tool steels of ledeburite type (high chromium and high speed steels) form a group of alloys produced by powder metallurgy by the gas atomization of melt followed by consolidation of rapidly solidified (RS) particles. They are characteristic by the high wear resistance caused by the carbide phases of solidification origin present in their microstructure.

Generally, the cooling and solidification of molten droplets in atomization process proceeds in several stages: rapid cooling, nucleation, rapid crystal growth during recalescence (quasi-adiabatic solidification), quasi-isothermal period of structure development in the semi-solidified particle and the cooling of solidified particle. In this paper, the nucleation temperature of molten rapidly cooled spherical particles from the Ch12MF4 tool steel in the gas atomization process is predicted on the base of classical theory of homogeneous nucleation. The influence of initial melt undercooling bellow the equilibrium liquidus temperature on the final microstructure development in the powder particles of different diameters is discussed.

2. Experimental materials and methods

For experimental measurements and theoretical study, two similar high alloyed tool steel of ledeburite type were applied, namely, the Ch12MF4 tool steel in the form of powder prepared by nitrogen gas atomisation and its equivalent K190 Isomatrix from BÖHLER Edelstahl GmbH company. The chemical composition of experimental steels is given in Table 1.

Table 1. Chemical composition of investigated steels.

Steel grade	Element [wt. %]					
	C	Cr	Mo	V	Si	Fe
Ch12MF4	2.37	12.06	1.2	4.0	-	bal.
K 190 Isomatrix	2.3	12.5	1.1	4.0	0.4	bal.

The process of quasi-equilibrium solidification of the Ch12MF4 steel was experimentally investigated using differential thermal analysis (DTA). In addition, material properties and solidification characteristics were computed by ThermoCalc and JMatPro software [14, 15]. According to DTA measurements and computed results, solidification of the Ch12MF4 steel in quasi-equilibrium conditions is initiated at the liquidus temperature of 1324 °C by the austenite crystallization followed by two eutectic reactions. The first eutectic reaction starts at the temperature T_{E1} = 1253 °C. During this eutectic reaction, the growth of two-phase mixture of austenite and MC carbides occurs. At the temperature T_{E2} = 1239 °C, the second eutectic reaction begins, which results in the formation of another two-phase mixture of austenite and M_7C_3 carbides. The process of quasi-equilibrium solidification is finished at the solidus temperature of 1225 °C. The phase identification was performed by transmission Mössbauer spectrometry and X-ray diffraction analysis [16-17].

Rapidly solidified (RS) powder particles of Ch12MF4 steel were screened into nine standard size granulometric fractions ranging from 50 μm to 56 μm up to more than 630 μm. The analysis of solidification microstructures developed in powder particles during non-equilibrium solidification was carried out using light microscopy (LM), scanning (SEM) and transmission electron microscopy (TEM).

To investigate the temperatures of spontaneous nucleation in the K190 Isomatrix alloy experimentally, the levitation techniques were exploited.

3. Solidification microstructures in rapidly solidified powders

A detailed analysis of solidification microstructures in 500 RS particles from each size fraction was performed [18]. The total number of 4500 particles was observed and analyzed. From the morphological point of view, three main types of solidification microstructures were identified RS powder particles (Figure 1): dendritic, cellular (grain-refined) and compound, representing the mixture of dendritic and cellular microstructures. The percentage population of the main types of solidification microstructures in analyzed granullometric fractions of RS powder is shown in Figure 2. Dendritic microstructure represents the dominant morphological variant in RS particles. It was found in each granullometric fraction while its percentage population decreased with the increase in particle size. On the other hand, the percentage population of compound and cellular (grain-refined) microstructures increases for the larger granullometric fractions. The considerable enhancement in percentage population of these microstructures was registered in the granullometric fractions larger than 125–168 μm. The appearance of cellular microstructures on a statistically significant scale was identified in the particles from the size fraction of 71-80 μm.

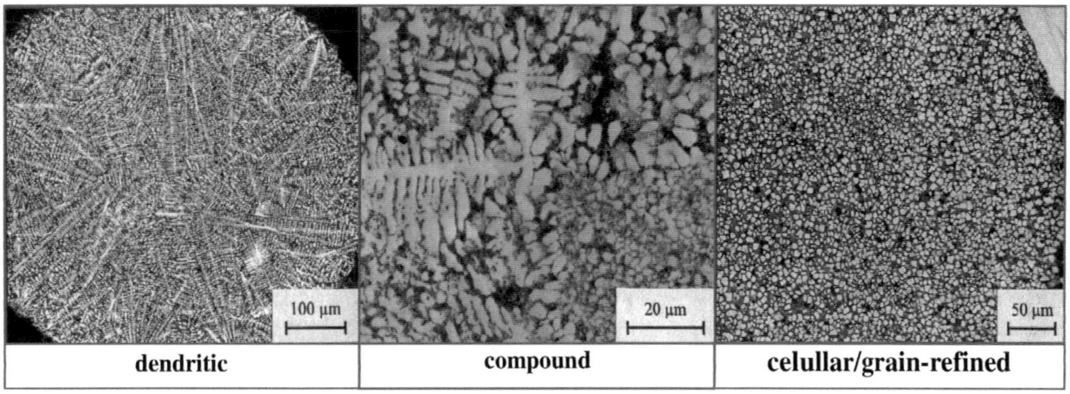

Figure 1. Main types of solidification microstructures identified in RS particles of the Ch12MF4 steel: dendritic, compound and cellular (grain refined).

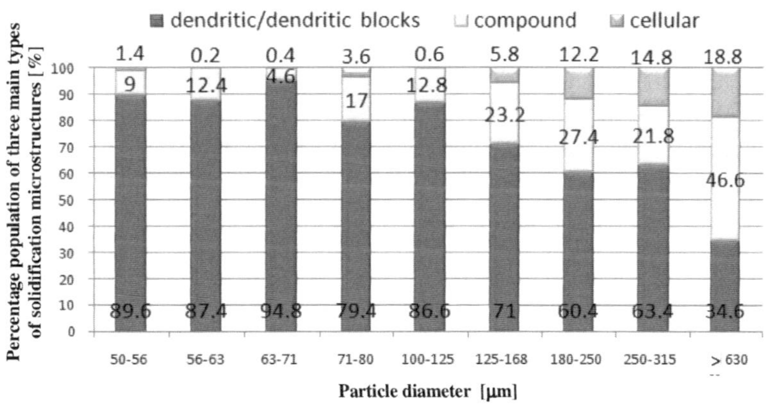

Figure 2. Percentage population of the main types of solidification microstructures identified in analyzed granullometric fractions of RS powder [18].

Based on the performed microstructural analysis it is supposed that the transition from dendritic to compound and cellular (grain-refined) microstructure occurs by the mechanism of thermally induced fragmentation of primary dendrites developed during the initial phase of solidification and following spheroidization of dendrite fragments during the quasi-isothermal period of solidification [17]. The time–temperature–microstructure model of the microstructure development in the rapidly solidified powder particles of Ch12MF4 steel was proposed [18] on the base of results of microstructural analyses and simulation model of the process of single droplet cooling and solidification during gas atomization. There was supposed in the simulation model that the process of droplet solidification starts from the multiple nucleation evens by the nucleation at the defined nucleation temperature from the interval from

IOP Conf. Series: Materials Science and Engineering **266** (2017) 012001 doi:10.1088/1757-899X/266/1/012001

1000 °C to 1300 °C corresponding to the initial melt undercooling below the equilibrium liquidus temperature from 324 K to 24 K. The influence of the nitrogen gas velocity, particle size and nucleation temperature on the thermal history of droplets during atomization was studied [18]. According to obtained results, the nucleation temperature has a crucial effect on the computed recalescence temperature and duration of quasi-isothermal period of solidification. From the microstructural point of view, these parameters are responsible for the formation of different morphological variants of solidification microstructures developed in RS powder particles of Ch12MF4 steel.

In this reason the following research was focused on the identification of the maximum possible undercooling before the nucleation, which can be attained in the droplets from Ch12MF4 steel with different diameters during gas atomization. Using the classical theory of homogeneous nucleation, the nucleation temperature of molten rapidly cooled spherical particles from this alloy with diameter from 40 μm to 600 μm in the gas atomization process was calculated using various estimations of parameters influencing the nucleation process. Moreover, the levitation experiments were conducted to measure the reachable undercooling in the melt of the alloy with similar chemical composition.

4. Mathematical model

According to the newtonian conditions, the thermal history of a spherical droplet can be described by the heat balance equation. The enthalpy decrease with time is equal to the heat extracted per time unit from the droplet surface to the surroundings [19]

$$-\frac{\mathrm{d}H}{\mathrm{d}t} = h(T - T_\mathrm{f})S \qquad (1)$$

The heat removed through the droplet surface S depends on the combine heat transfer coefficient h and the temperature of atomizing gas T_f. Taking into account the enthalpy change of liquid and solid phases and the latent heat released, the time variation of the droplet temperature becomes

$$\frac{\mathrm{d}T}{\mathrm{d}t} = \left[-\frac{6h(T - T_\mathrm{f})}{d} + \rho_\mathrm{L} l_\mathrm{t} \frac{\mathrm{d}f_\mathrm{S}}{\mathrm{d}t}\right]\left[\rho_\mathrm{S} c_\mathrm{C} f_\mathrm{S} + \rho_\mathrm{L} c_\mathrm{L}(1 - f_\mathrm{S})\right]^{-1} \qquad (2)$$

where d is the droplet diameter, ρ and c are the density and specific heat of solid (index S) and liquid (index L) phases, respectively, l_t is the latent heat and f_S is the solid fraction.

Solution of the equation (2) requires definition of geometrical, physical, initial and boundary conditions. The crucial problem consists in determination of cooling conditions, i. e. evaluation of the temperature and time dependent heat transfer coefficient during the droplet cooling by mechanisms of convection and radiation, determination of nucleation temperature and the velocity of solid-liquid (S-L) interface in the dependence on the melt undercooling.

Supposing the heat extraction from the droplet surface by convection and radiation, the boundary condition of the 3rd type [19] can be written in the form

$$-\lambda(grad\,T)_\mathrm{w} = h(T_\mathrm{w} - T_\mathrm{f}) \qquad (3)$$

where T_w is the droplet surface temperature and λ is the thermal conductivity. Considering the energy losses to the surroundings characterized by the coefficient γ, the gas temperature can be determined from the equation

$$T_\mathrm{f} = T_\mathrm{f0} + \int_{T_0}^{\mathrm{T}} (1 - \gamma)\frac{\dot{m}c}{\dot{m}_\mathrm{f} c_\mathrm{f}} \mathrm{d}T \qquad (4)$$

where $\dot{m}/\dot{m}_\mathrm{f}$ is the melt to gas mass flow ratio, c and c_f are specific heat capacities of alloy and gas, respectively and T_f0 is the gas initial temperature.

The combined heat transfer coefficient h represents the sum of the convection heat transfer coefficient h_K and the equivalent radiation heat transfer coefficient h_R, $h = h_K + h_R$. Using the Ranz-Marshall criterial equation [19], the heat transfer coefficient can be determined from the relationship

$$h = h_K + h_R = \frac{\lambda_f}{d}\left[2 + 0.6\, Re^{1/2}\, Pr^{1/3}\right] + \varepsilon\sigma_0\left(T^2 + T_f^2\right)\left(T + T_f\right) \tag{5}$$

where Re is the Reynolds number, Pr is the Prandtl number, λ_f is the gas thermal conductivity, ε is the droplet emissivity and σ_0 represents the Stefan-Boltzmann constant.

The convection heat transfer coefficient depends particularly on the Reynolds number and relative velocity of the gas and droplet [20].

In order to determine the maximal possible droplet undercooling prior to solidification and to study the limiting conditions of droplet crystallization, homogeneous nucleation in the undercooled droplets is assumed. Nucleation occurs at the temperature T_N in the moment when one stable nucleus of critical size is created in the molten droplet with the volume V [7, 8]

$$N = V\int_{T_L}^{T_N} \frac{J_v(T)\, dT}{\dot{T}} = 1\,. \tag{6}$$

The symbol $\dot{T}$ denotes the cooling rate and $J_v(T)$ is the steady-state volumetric nucleation rate given by the relationship

$$J_v = \Omega_v\, exp\left(\frac{\Delta G^*}{k_B T}\right) \tag{7}$$

where ΔG^* is the activation barrier for nucleation, k_B is the Boltzmann constant and Ω_v is the preexponential factor with the value of 10^{35} cm^{-3}s^{-1} suggested by Perepezko [21] or $\Omega_v = 10^{30}/\eta$ [cm^{-3}s^{-1}] according to Thurnbull [22]. In this case Ω_v is the function of dynamic viscosity which can be for undercooled melts estimated from the relationship [23]

$$\eta = 10^{-3.3}\, exp\left[3.34 T_L/\left(T - T_g\right)\right] \qquad \text{[poise]} \tag{8}$$

Supposing formation of a spherical nucleus, the activation energy ΔG^* can be calculated as [61]

$$\Delta G^* = \frac{16}{3}\frac{\pi\sigma^3}{\Delta G_v^2} \tag{9}$$

where σ is the solid/liquid interfacial energy and ΔG_v is the difference of Gibbs free enthalpy of liquid and solid phases per unit volume. Both of these parameters have significant influence on calculated nucleation temperatures. Several authors have tried to estimate them by different mathematical approximations [25]. In this work, the interfacial energy is calculated using the relationship introduced by Thompson and Spaepen [23]

$$\sigma = \frac{\alpha \Delta S_m T}{\left(N_A V_m^2\right)^{1/3}} \tag{10}$$

where α is a structure dependent constant ($\alpha = 0.71$ for b.c.c. and $\alpha = 0.86$ for f.c.c./h.c.p.), ΔS_m is the molar entropy of fusion, N_A is the Avogadro number and V_m the molar volume.

The Gibbs free energy difference between solid and liquid phases ΔG_v, is given by the relationship

$$\Delta G_v = \frac{\Delta H_v \Delta T}{T_L} - \int_T^{T_L} \Delta c_p dT + T\int_T^{T_L} \frac{\Delta c_p}{T} dT \tag{11}$$

in which ΔH_v is the volumetric enthalpy of fusion (volumetric latent heat), Δc_p is the heat capacity difference between undercooled liquid and solid phases and $\Delta T = T_L - T$ is the melt undercooling.

As the experimental values of ΔG_v and Δc_p, respectively, are not available, their temperature dependences are approximated assuming various forms of Δc_p. For example, Thurnbull [10] proposed $\Delta c_p = 0$ for pure metals leading to the linear dependence of ΔG_v on the melt undercooling. Selected approximations for the computation of Gibbs free energy difference between solid and liquid phases ΔG_v are summarized in Table 2. All considered expressions for ΔG_v are related to the linear approximation of Turnbull through some correction factors.

Table 2. Selected approximations for the computation of Gibbs free energy difference ΔG_v.

Author(s)	Approximation of ΔG_v	Reference
Turnbull	$\Delta G_v = \frac{\Delta H_v \Delta T}{T_L}$	[26]
Hoffman	$\Delta G_v = \frac{\Delta H_v \Delta T}{T_L}\left(\frac{T}{T_L}\right)$	[27]
Thompson and Spaepen	$\Delta G_v = \frac{\Delta H_v \Delta T}{T_L}\left(\frac{2T}{T_L + T}\right)$	[28]
Battezzati and Garrone	$\Delta G_v = \frac{\Delta H_v \Delta T}{T_L} - 0.8\frac{\Delta H_v}{T_L}\left(\Delta T - T\, ln\frac{T_L}{T}\right)$	[29]
Jones and Chadwick	$\Delta G_v = \frac{\Delta H_v \Delta T}{T_L} - \frac{\Delta c_p^L (\Delta T)^2}{(T_L + T)}$	[30]
Dubey and Ramachandrarao	$\Delta G_v = \frac{\Delta H_v \Delta T}{T_L} - \frac{\Delta c_p^L (\Delta T)^2}{2T}\left(1 - \frac{\Delta T}{6T}\right)$	[31]
Singh a Holz	$\Delta G_v = \frac{\Delta H_v \Delta T}{T_L}\frac{7T}{T_L + 6T}$	[32]

Δc_p^L is the difference of thermal capacity between solid and liquid phases at the liquidus temperature

5. Results and discussion

The theory of homogenous nucleation was applied to predict achievable levels of undercooling of the spherical droplets from Ch12MF4 steel with diameters from 20 μm to 600 μm in the process of nitrogen gas atomization. The thermal properties of the Ch12MF4 steel in the dependence on temperature (Figure 3) were defined on the base of their computation by JMatPro software [15] and dilatometry, differential thermal analysis (DTA) and laser flash (LFA) measurements.

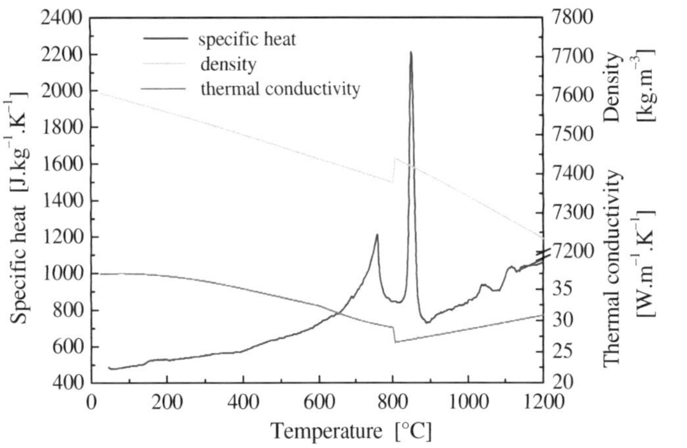

Figure 3. Thermal properties of the Ch12MF4 steel in the dependence on temperature.

The estimation of nucleation temperature T_N based on the Eq. (6) derived from the assumption that the droplet solidification started in the moment when one nucleus of critical size had been formed. According to this equation, the nucleation temperature depends on the cooling rate and the steady-state volumetric nucleation rate. The cooling rate is strongly dependent on the droplet diameter and the heat transfer coefficient governed by the relative velocity of a droplet and atomizing gas [20]. In Figure 4, the maximum cooling rate is plotted as a function of droplet diameter for chosen initial gas velocities representing the different possible positions of a droplet in atomizing conus. The smaller droplets are cooled more intensive thus they can reach deeper undercoolings below the equilibrium liquidus temperature prior to nucleation in comparison with larger droplets.

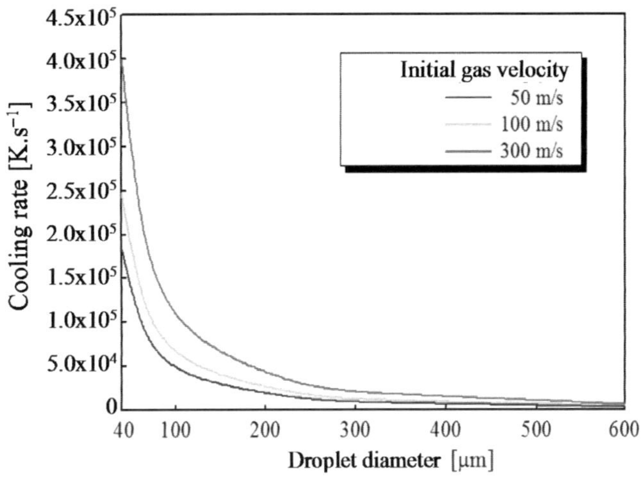

Figure 4. Maximum cooling rate *v.s.* the droplet diameter for different initial velocities of atomizing gas.

The steady-state volumetric nucleation rate is influenced particularly by the solid/liquid interfacial energy, dynamic viscosity and the Gibbs free energy difference between solid and liquid phases. The dependence of the solid/liquid interfacial energy and the viscosity on the melt undercooling is shown in Figure 5. The viscosity enhances by several orders with the increase in droplet undercooling. On the other hand the surface tension for deeper undercoolings decreases.

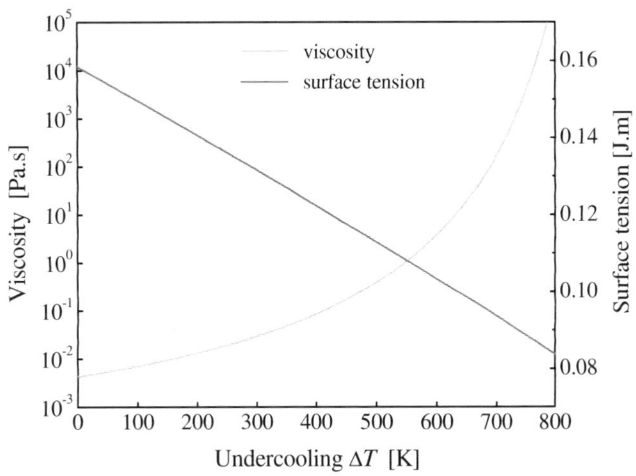

Figure 5. Dependence of the dynamic viscosity and the surface tension on the droplet undercooling.

The dependence of the Gibbs free energy difference ΔG_v computed according the approximations introduced in Table 2 on the droplet undercooling is depicted in Figure 6. As it follows from this figure, for the small undercoolings (up to approximately 50 K), considered relationships results in similar values of ΔG_v. However, the deviations of the Gibbs free energy differences determined according to selected approximations are getting to increase apparently for higher undercoolings. The simple formula suggested by Turnbull provides the upper estimation for the value of the Gibbs free energy difference between solid and liquid phases. The lowest values of ΔG_v were computed applying the Dubey and Ramachandrarao approximation proposed particularly for the glass forming alloys and organic liquids.

In Figure 7, computed temperatures of homogeneous nucleation are plotted as a function of droplet diameter. The maximal nucleation temperature (or in other words the minimal initial undercooling) was calculated for all investigated diameters of droplets using the Turnbull approximation of ΔG_v.

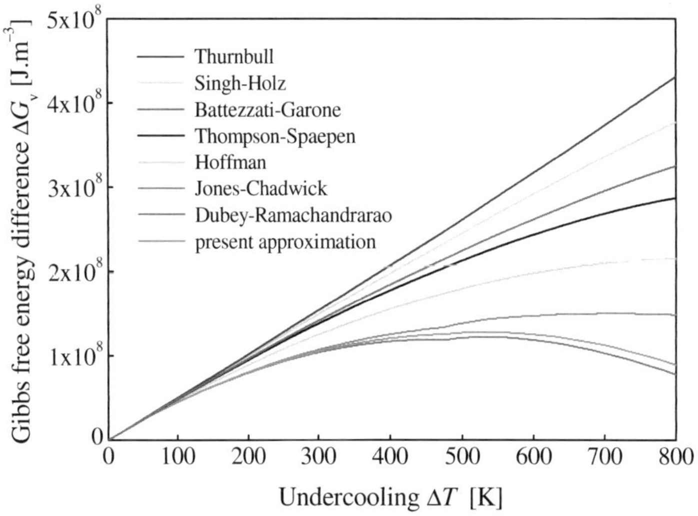

Figure 6. Dependence of the Gibbs free energy difference ΔG_v computed according the approximations introduced in Table 2 on the melt undercooling.

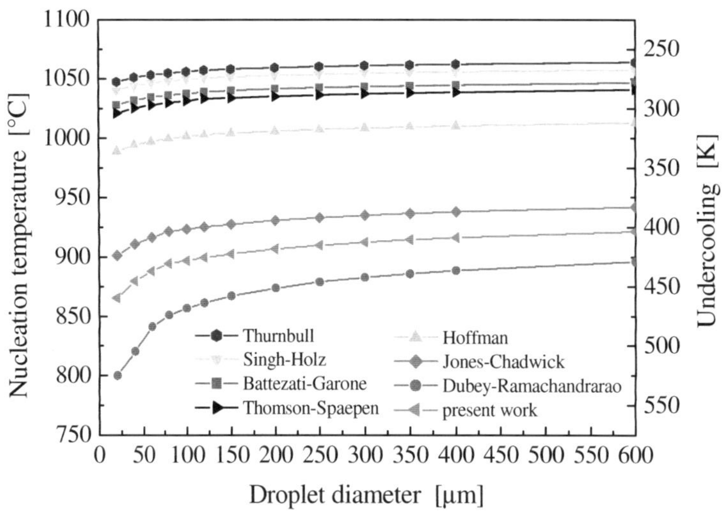

Figure 7. Dependence of the nucleation temperature and initial undercooling on the droplet diameter.

The lowest values of nucleation temperature were provided by the Dubey and Ramachandrarao relationship. In addition, in this case the droplet diameter has higher influence on the computed nucleation temperature. The differences between the nucleation temperatures predicted for droplets with the diameter of 20 μm and 600 μm are larger comparing with those computed using Turnbull formula. Nucleation temperatures computed in the framework of our research by direct application of the Eq. (11) are closer to bottom limit suggested by Dubey and Ramachandrarao. The upper and bottom limits of nucleation temperature and initial undercooling together with values computed by presented model are summarized in Table 3. For the droplets with the largest considered diameter of 600 μm, the initial undercoolings prior to nucleation can reach the values from 260 K (Turnbull approximation) to 428 K (Dubey-Ramachandrarao approximation) corresponding to the nucleation temperatures of 1064 °C and 896 °C, respectively.

Table 3. Nucleation temperature T_N and initial undercooling ΔT_N computed using different approximations of ΔG_v for selected diameters of droplets.

Approximation of ΔG_v	**Droplet diameter**							
	20 μm		**80 μm**		**120 μm**		**600 μm**	
	T_N [°C]	ΔT_N [K]	T_N [°C]	ΔT_N [K]	T_N [°C]	ΔT_N [K]	T_N [°C]	ΔT_N [K]
Turnbull (upper limit)	1047	276	1055	269	1057	267	1064	260
Dubey (bottom limit)	800	524	851	473	861	463	896	428
Current work	865	459	895	429	899	425	921	403

In compliance with the time–temperature–microstructure model of the microstructure development in the rapidly solidified powder particles from Ch12MF4 steel [18], the microstructure transition from dendritic to compound and cellular can occurs if both from following conditions are fulfilled:

- the recalescence temperature is higher then the equilibrium temperature of the eutectic reaction T_{E1}=1253 °C and
- the duration of the quasi-isothermal plateau above the T_{E1} temperature is sufficiently long.

Figure 8 illustrates the thermal history of a droplet with the diameter of 80 μm rapidly solidifying from different supposed nucleation temperatures and conditions of cooling by nitrogen with the initial velocities of 50 m.s^{-1} and 300 m.s^{-1}, respectively.

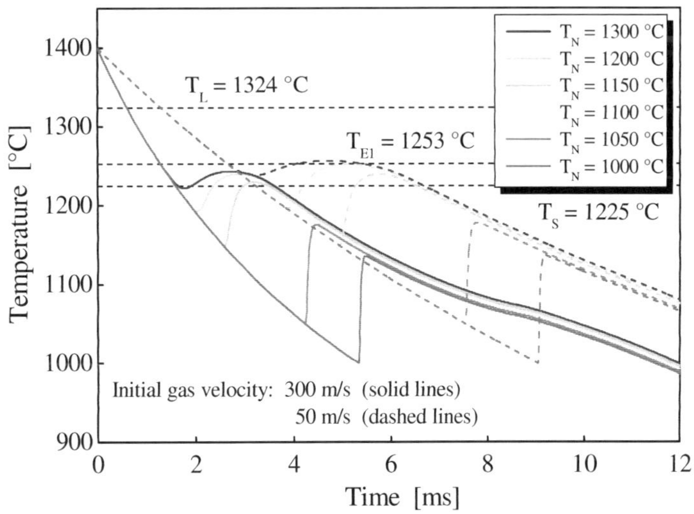

Figure 8. Thermal history of a droplet with the diameter of 80 μm rapidly solidifying from different supposed nucleation temperatures T_N.

If the initial gas temperature is 300 m.s^{-1}, the recalescence temperature does not reach the equilibrium temperature of the eutectic reaction T_{E1}=1253 °C for any assumed nucleation temperature. However, in case of lower initial gas velocity 50 m.s^{-1} and nucleation temperatures above approximately 1200 °C, the first condition for microstructure transition is attained. This finding corresponds to results of microstructural analysis and to the increase in percentage population of cellular and mainly compound microstructures presented in Figure 2.

Figures 9 and 10 show in more details the influence of nucleation temperature and the initial gas velocity on the recalescence temperature and on the duration of the phase of quasi-isothermal solidification for droplets with the diameter of 40 μm, 80 μm, 120 μm and 600 μm.

For the smallest droplet with the diameter of 40 μm, the maximum recalescence temperature attain values bellow or only around the equilibrium solidus temperature (Figure 9). In this reason, the thermally induced fragmentation of dendrites is improbable. The temperature of droplets with the diameter of 80 μm reach the temperature T_{E1} for the nucleation temperatures higher than 1220 °C only in the case of less intensive cooling conditions (initial gas velocity of 50 m.s^{-1}). For these nucleation temperatures, the

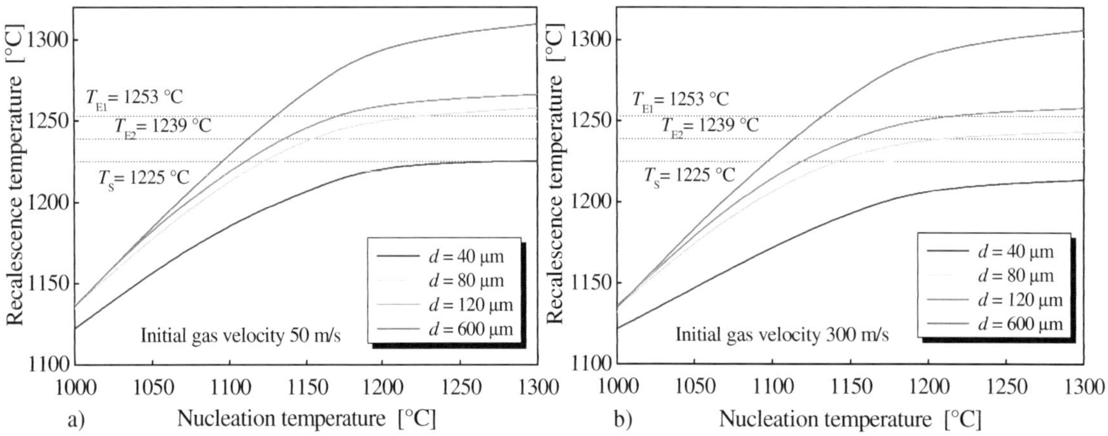

Figure 9. Dependence of recalescence temperature on the nucleation temperature for chosen particle diameters and the initial gas velocity of a) 50 m.s^{-1} and b) 300 m.s^{-1}.

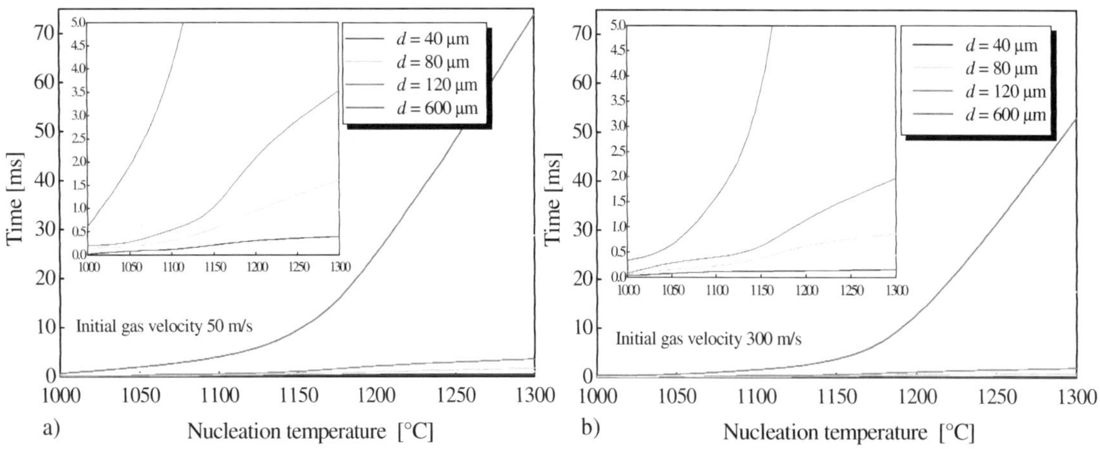

Figure 10. Dependence of the duration of quasi-isothermal plateau on the nucleation temperature for chosen particle diameters and the initial gas velocity of a) 50 m.s^{-1} and b) 300 m.s^{-1}.

duration of quasi-isothermal plateau is more than 1.1 ms. Temperatures in droplets with the diameter of 120 μm rise above the temperature T_{E1} also for the initial gas velocity of 300 m.s^{-1} if the nucleation temperature is higher than 1213 °C or for the nucleation temperatures above 1170 °C when the initial gas velocity is 50 m.s^{-1}. Duration of the quasi-isothermal plateau is 1.2 ms and 1.5 ms, respectively. As the population of compound and cellular microstructures significantly increases in the powder particles larger than 120 μm in diameter (Figure 2), it can be supposed that above mentioned conditions can be sufficient for the initiation of thermally induced fragmentation dendrite and spheroidization of dendrite fragments.

The highest recalescence temperatures and the longest periods of quasi-isothermal solidification are attained in the largest droplets. However, the recalescence temperatures of droplets with the diameter of 600 μm are below the solidus temperature if the nucleation temperatures are lower than approximately 1095 °C. For the nucleation temperatures less than 1130 °C, the recalescence temperatures do not reach the temperature of the eutectic reaction T_{E1}. Under these conditions, the austenite dendrites developed in the phase of recalescence remain unchanged (thermally not influenced) and the final microstructure of solidified particle is dendritic. According computed temperatures of homogeneous nucleation (Table 3), also the largest molten droplets from Ch12MF4 steel can be undercooled below the temperature of 1130 °C and even below the temperature of 1095 °C. This conclusion can be confirmed by the fact that in largest granullometric fractions of RS powder, the percentage population of dendritic microstructure was found to be 34.6 % (Figure 2).

Finally, the computed results were compared with the nucleation temperatures measured in the samples from the tool steel K190 Isomatrix during levitation experiments in the Leibniz Institute for Solid State and Materials Research in Dresden [33-34]. Together 11 samples with the diameters from 6 mm to 8 mm were prepared and treated during 59 experiments. Samples were heated and melted in the vacuum chamber and then cooled by inert gas in electromagnetic levitation or splat quenched on Cu, Sn-Cu or Pb-Cu substrates. The temperatures of spontaneous nucleation along with recalescence temperatures were measured. The melt undercoolings from 50 °C to 290 °C below the equilibrium liquidus temperature were achieved (Figure 11). The spontaneous solidification of samples started the most frequently at undercoolings from 200 °C to 250 °C. Computed results are in a good agreement with the experiments. Measured levels of undercooling correspond approximately to the upper limit of undercooling computed according to Turnbull formula. Taking into account the smaller volumes of droplets and the higher cooling rates acting during gas atomization process, the lower temperatures of homogeneous nucleation computed by the present model or other considered approximations (Table 2) can represent very probable estimation of actual levels of droplet undercooling.

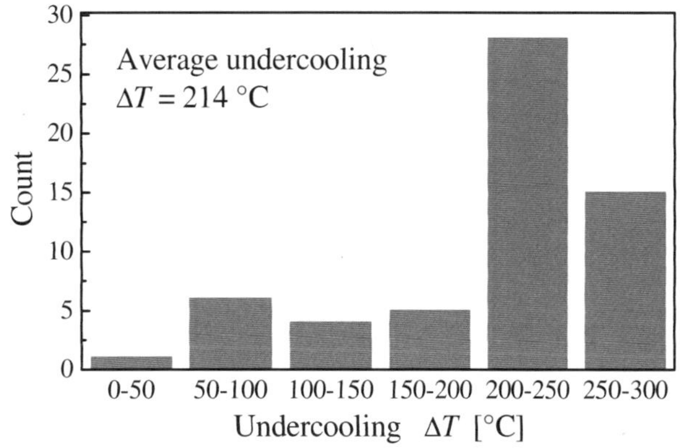

Figure 11. Histogram of achieved undercoolings prior to spontaneous nucleation in levitation experiments [33].

6. Conclusions

Based on the microstructural analysis, computational results and levitation experiments, the following conclusions regarding nucleation and microstructure development during rapid solidification of undercooled droplets from the Ch12MF4 tool steel in the nitrogen gas atomization can be made:

- According to the morphology of austenite, three main types of solidification microstructures were identified in RS powder particles from Ch12MF4 steel - dendritic, compound and cellular (grain-refined).
- Dendritic microstructure was observed in all investigated granullometric fractions of RS powder and its percentage population decreased for larger particles. Statistically significant appearance of compound and cellular microstructures was identified in granullometric fraction from 71-80 μm but considerable enhancement of their proportion was from the granullometric fraction from 125-168 μm.
- It is supposed that the transition from dendritic to compound and cellular microstructure occurs by the mechanism of thermally induced fragmentation of austenite dendrites developed in the phase of recalescence during the quasi-isothermal period of solidification.
- Microstructure transition can occur when the recalescence temperature is higher than both eutectic temperatures and the duration of quasi-isothermal plateau is sufficiently long. Both these parameters are significantly influenced by nucleation temperature.
- Attainable undercooling of molten droplets during gas atomization was predicted using the theory of homogeneous nucleation at the level from 260 K to 524 K.
- During levitation experiments, the most frequently the undercoolings from the interval from 200 K to 250 K were achieved.
- According to computed results, the conditions for microstructure transitions can be fulfilled for particles larger than 80 μm in diameter and mainly for particles with diameter more than 120 μm.
- Dendritic microstructure in largest particles with diameter of 600 μm can remain unchanged if the nucleation temperature is below 1130 °C.

Acknowledgement

The research has been supported by the Scientific Grant Agency of the Slovak Republic (VEGA) within the project No. 1/0811/14, the project No. 1/1010/16 and the project ITMS 26220120048 under the Research & Development Operational Programme funded by the ERDF.

References

[1] Herlach D M 1994 Non-equilibrium solidification of undercooled metallic melts *Mat Sci Eng R* **2** 4–5 177-272.
[2] Rafii-Tabar H and Chirazi A 2002 Multi-scale computational modelling of solidification phenomena *Phys Rep* **365** 3 145-249.
[3] Mondal K and Murty B S 2006 Prediction of maximum homogeneous nucleation temperature for crystallization of metallic glasses *J Non-Cryst Solids* **352** 50–51 5257-5264.
[4] Jia F, Zhao D and Wang M 2016 Selective nucleation and self-organized crystallization *Prog Cryst Growth* **62** 2 252-272.
[5] Perepezko J H and Wilde G 2016 Melt undercooling and nucleation kinetics *Curr Opin Solid State Mater Sci* **20** 1 3-12.
[6] Karma A and Tourret D 2016 Atomistic to continuum modeling of solidification microstructures *Curr Opin Solid State Mater Sci* **20** 1 25-36.

[7] Jacobson L A and McKittrick I 1994 Rapid solidification processing *Mater Sci Eng* **R11** 8 355-408.

[8] Herlach D, Galenko P and Holand-Moritz D 2007 *Metastable solids from undercooled melts* Pergamon Press Amsterdam ISBN 978-0-08-04368-8

[9] Herlach D M 2014 Non-Equilibrium Solidification of Undercooled Metallic Melts *Metals* **4** 2 196-234 doi:10.3390/met4020196

[10] Perepezko J H, Hockel P G and Paik J S 2002 *Termochim Acta* **388** 129-141

[11] Perepezko J H and Hildal K 2008 Metallic glass formation reactions and interfaces *Mater Sci Eng B* **148** 171-178

[12] Perepezko J H 2005 Nucleation controlled phase selection during solidification *Mater Sci Eng A* **413–414** 389-397

[13] Clavaguera-Mora M T et al 2002 *Prog Mater Sci* **47** 559-618

[14] Thermo-Calc Version 2017a. http://www.thermocalc.com

[15] JMatPro, Sente Software Ltd., Surrey Technology Centre, Surrey GU2 7YG, United Kingdom.

[16] Behúlová M, Moravčík R, Kusý L, Čaplovič Ľ, Grgač P and Stanček L 2001 Influence of atomisation on solidification microstructures in the rapidly solidified powder of the Cr-Mo-V tool steel *Mater Sci Eng* **A304-306** 540-543

[17] Grgač P, Behúlová M, Moravčík R and Mesárošová J 2012 *Mater Research* **15** 705-712.

[18] Behúlová M, Mesárošová J and Grgač P 2014 Analysis of the influence of the gas velocity, particle size and nucleation temperature on the thermal history and microstructure development in the tool steel during atomization *J Alloys Compd* **615** 217-223.

[19] Incropera F P, Dewitt D P 1996 *Fundamentals of heat and mass transfer* J Wiley and Sons New York, 1996.

[20] Behúlová M 2005 Influence of initial nitrogen gas velocity on the microstructure development od Ch3F12 alloy in atomization process *Metallic materials* **43** 2 145-157

[21] Perepezko J H 1985 In Proc. of NATO Adv. Research Workshop on RS Technologies, Theuern

[22] Turnbull D 1952 Kinetics of Solidification of Supercooled Liquid Mercury Droplets *J Chem Phys* **20** 411

[23] Thompson C V and Spaepen F 1983 Homogeneous crystal nucleation in binary metallic melts *Acta metall* **31** 12 2021-2027

[24] Turnbull D 1950 Phase Changes *J Appl Phys* **21**1022

[25] Mondal K, Kumar A, Gupta G and Murty B S 2009 Temperature and structure dependency of solid–liquid interfacial energy *Acta Mater* **57** 3422–3430

[26] Turnbull D 1956 *Solid St Phys* Academic Press New York **3** 225

[27] Hoffman J D 1958 Thermodynamic Driving Force in Nucleation and Growth Processes *J Chem Phys* **29** 1192

[28] Thompson C V and Spaepen F 1979 On the approximation of the free energy change on crystallization *Acta Metall* **27** 1855

[29] Battezzati L and Garrone E 1984 *Z Metallkunde* **75** 305

[30] Jones D R H and Chadwick G A 1971 An expression for the free energy of fusion in the homogeneous nucleation of solid from pure melts *Phil Mag* **24** 995

[31] Dubey K S and Ramachandrarao P 1984 On the free energy change accompanying crystallisation of undercooled melts *Acta Metall* **32** 91

[32] Singh H B and Holz A 1983 Stability limit of supercooled liquids *Solid State Communications* **45** 985

[33] Grgač P, Lipták M, Behúlová M, Čaplovič Ľ, Lindenkreuz H-G and Löser W 2007 Influence of melt undercooling on the microstructure of levitated Cr–Mo–V tool steel *Mat Sci Eng A* **449-451** 658-661

[34] Behúlová M, Lipták M, Grgač P, Löser W amd Lindenkreuz H-G 2009 Comparison of microstructures developed during solidification of undercooled tool steel in levitation and on a substrate *Journal of Physics: Conference Series* **144** 1-4

IOP Conf. Series: Materials Science and Engineering **266** (2017) 012002 doi:10.1088/1757-899X/266/1/012002

Contribution to the aluminum–tin–zinc ternary system

J Drápala, G Kostiuková and M Losertová

Vysoká škola báňská – Technical University of Ostrava, Faculty of Metallurgy and Materials Engineering, 17. listopadu 15, 708 33 Ostrava – Poruba, Czech Republic

Jaromir.Drapala@vsb.cz

Abstract. The Sn–Zn–Al alloys are one of significant candidates in the proposal of alternative lead-free solders for higher temperature soldering. This paper deals with the study of the aluminum–tin–zinc system. Twenty Sn–Zn–Al alloys together with six binary Sn–Zn alloys were prepared and studied experimentally. Alloys were prepared from pure Sn, Zn and Al (melting and cooling in a vacuum resistance furnace). The specimens were studied metallographically including the micro-hardness measurements, complete chemical analysis (ICP-AES, OES), X-ray micro-analysis of alloys by SEM and EDX in order to determine the composition and identification of individual phases. Significant temperatures and enthalpies of phase transformations were determined by DTA. After long-term annealing of selected alloys in vacuum followed by quenching the structural and chemical microanalyses of the present phases and their limit concentrations were carried out. The achieved results were compared with the thermodynamic modelling of the ternary Sn–Zn–Al system (computer programs THERMOCALC, MTDATA, PANDAT and databases CALPHAD, COST). Electrical resistivity, density, magnetic susceptibility and wettability of Sn–Zn–Al solders were measured as well.

1. Introduction

The Sn–Zn–Al ternary system can be one of applicable alternative alloys to high lead content Pb–Sn solders, when the soldering process reaches temperatures as high as 300 °C. These solders can be used above all in automotive industry, in heat exchangers and in special applications. Tin – zinc alloys are used as solders for aluminum.

The tin – zinc binary system diagram is of a eutectic type. The eutectic reaction proceeds at the temperature of 198.5 °C and the concentration of 85.1 at.% Sn. The zinc limit solubility in tin at this temperature corresponds to 0.6 at.% Zn, while on the other hand, the tin maximum solubility in hexagonal zinc is only 0.039 at.% Sn [1].

The binary diagram of the aluminum – tin system [1, 2] is also of a eutectic type with a considerably limited solubility in the area of solid solution (Al) and tetragonal (Sn). The tin solubility in (Al) solid solution exhibits a retrograde character with a maximum of approximately 0.026 at.% Sn at the temperature of 625±20 °C [2].

The Al–Zn equilibrium phase diagram [1] represents a eutectic system involving another reaction and the existence of a miscibility gap area in solid phase α (Al).The eutectic reaction proceeds at 381 °C closer to Zn side (the eutectic point corresponds to 88.7 at.% Zn). (Al) solid solution has an extended area of homogeneity interrupted at the temperature of 351.5 °C and the content of 39.5 at.% Zn. Between this temperature and the temperature of 277 °C, (Al) disintegration into two solid solutions occurs: (Al) with a lower Zn content and (Al") with a high zinc content (up to 59 at.%). The

Content from this work may be used under the terms of the Creative Commons Attribution 3.0 licence. Any further distribution of this work must maintain attribution to the author(s) and the title of the work, journal citation and DOI.
Published under licence by IOP Publishing Ltd

Zn solubility in (α Al) increases from 2.2 at.% at 110 °C to 16.5 at.% at 277 °C. Above temperature 277 °C the zinc solubility increases from 59±1 at.% Zn at 277 °C to 67±1 at.% Zn at the eutectic reaction. The maximum solubility of Al in (Zn) is 2.8±0.2 at.% Al at the eutectic temperature and decreases along with a decreasing temperature to 1.6 at.% Al at 277 °C and down to 0.07 at.% Al at the room temperature. Then, (Al) + (Zn) two-phase structure can be expected in the equilibrium crystallization of Al–Zn alloys. In Table 1 thermodynamic values and reactions in Al–Sn, Sn–Zn and Al–Zn binary systems are listed.

Table 1. Thermodynamic values of phase transformations of Sn, Zn and Al pure metals and equilibrium compositions of solidus and liquidus at the eutectic or eutectoid reaction in Al–Sn, Sn–Zn and Al–Zn binary systems [1, 2].

Metal A	T_m^A (°C)	ΔH_m^A (J/mol)
Sn	232	7029
Zn	419.6	7322
Al	660.5	10711

T_m^A – melting point of metal A
ΔH_m^A – molar enthalpy of fusion for metal A
T_E – eutectic temperature
X_{SE} – maximal solubility of element B in A at T_E
X_{LE}– concentration of element B in liquidus at T_E
* – eutectoid reaction

Binary system	X_{SE} (at.%)	X_{LE} (at.%)	T_E (°C)
Sn–Al	1.0	2.4	228
Al–Sn	<0.01	97.6	228
Sn–Zn	0.6	14.9	198.5
Zn–Sn	0.04	85.1	198.5
Zn–Al	2.8	11.3	381
Al–Zn	67.0	88.7	381
Zn–Al*	1.6*	41*	277*
Al–Zn*	16.5*	59*	277*

The Al–Sn–Zn ternary system was studied in details by Prowans [3] and later by Nayak [4] and other authors [5, 6]. A revision of the Al–Sn–Zn ternary system and a thermodynamic calculation was performed by Fries [7] (see Figure 1). The latest findings on this system are mentioned on www.msiport.com [8] and www1.asminternational.org [9] proposed by M. Hubert-Protopopescu [10]. When researching this system it was found out that the liquidus areas in the Al–Sn–Zn ternary system including their isotherms differ according to data of authors [3-10] both in a position of the eutectic valley and in a character of the isoliquidus courses. So, there are significant disproportions in describing this system according to various authors. The thermodynamically shaped liquidus area according to Fries [7] is closer to the diagram designed experimen-tally by Nayak [4]. In Table 2 are summarized the invariant equilibria in the Al–Sn–Zn ternary system.

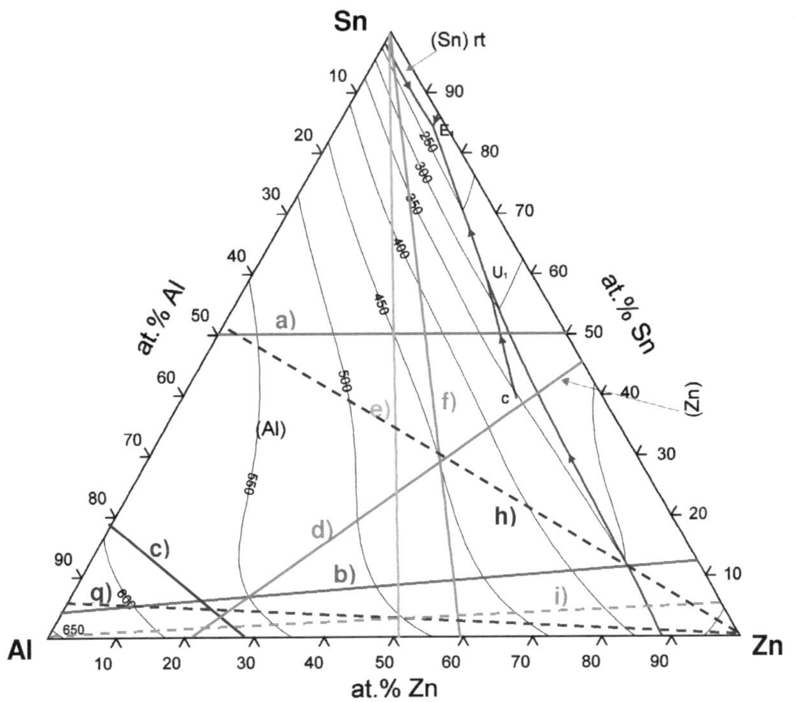

Figure 1. Thermodynamic calculated liquidus surface of the Al–Sn–Zn ternary system by [7, 10] with the denotation of nine sections forimage of isopleths.

Table 2. Thermodynamic invariant equilibria in the Al–Sn–Zn ternary system.

U_1	L + (Al) ↔ (Al)" + (Zn)					E_1	L ↔ (Al) + (Zn) + (Sn)				
T (°C)	Phase	Sn (at.%)	Zn (at.%)	Al (at.%)	Ref.	*T* (°C)	Phase	Sn (at.%)	Zn (at.%)	Al (at.%)	Ref.
277.8	L	61.76	31.79	6.45	[7]	195.4	L	84	13.86	2.14	[7]
	(Al)	0.01	14.17	85.82			(Al)	0	5.59	94.41	
	(Al)"	0.02	58.97	41.01			(Zn)	0.05	99.35	0.6	
	(Zn)	0.13	98.24	1.63			(Sn)	98.74	0.61	0.65	
						196	L	86.6	10.7	3.1	[5]
						196	L	77.7	16.6	5.7	[6]
						197.3	L	84.7	12.9	2.4	[3]
						198	L	82.5	15	2.5	[4]

On the basis of structural analyses, SEM and x-ray analysis EDX and DTA, the aim of this paper was to examine validity of the published diagrams, to perform thermodynamic calculations of phase equilibria after long-term isothermic annealing of specimens of selected alloys. Other properties of the selected Al–Sn–Zn alloys, such as resistivity, density, magnetic susceptibility and wettability are tested as well.

2. Assessment of the Al–Sn–Zn ternary system

Figure 2 shows isothermal cross-sections through the ternary Al–Sn–Zn system at the temperatures of 300 °C to 270 °C according to [7-11]. The diagram in Figure 2 b) is beside the character of the existence of (Al) + (Al") phases area, which penetrate deep inside the diagram at the expense of three-phase (Al) + (Al") + L area in the cross-section at 290 °C. Comparing the cross-sections at the temperatures of 300 °C and 280 °C, it is obvious that the cross-section at 290 °C is improperly plotted in term of the coexistence of phases. Sebaoun [11] studied the Al–Sn–Zn phase diagram using the reactive diffusion with preparing three diffusion pairs: Zn-10 wt.% Sn/Al, Zn-28 wt.% Al/Sn and Al-10 wt.% Sn/Zn (see lines **g**, **h** and **i** in Figure 1), which were annealed for a period of 3 up to 6 hours at the temperatures of 270, 280 and 290 °C. Following annealing, the specimens were water-quenched rapidly. From structural and chemical microanalyses (SEM/EDX) the respective diffusion routes were found out, on the basis of which a modified Al–Sn–Zn diagram was created (see the isothermal cross-sections in Figure 2 b), c) and d).The following phases were found out: (Al), (Al"), (Zn) and a solidified tin-rich melt. In all cases the tin solubility in (Al") phase was < 0.8 at.% Sn. That's why the isothermal cross-section in Figure 2 b) has been explicated incorrectly.

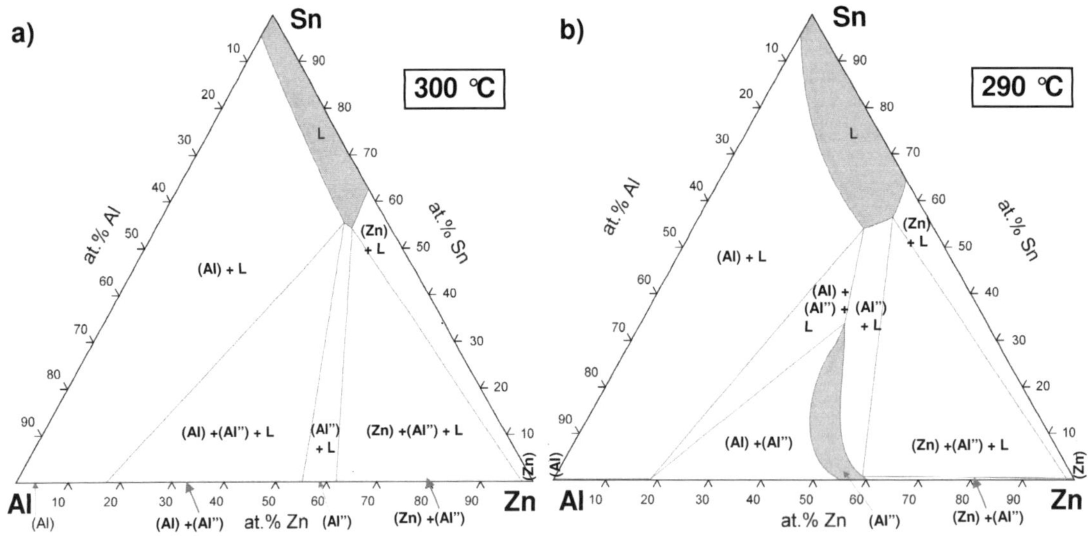

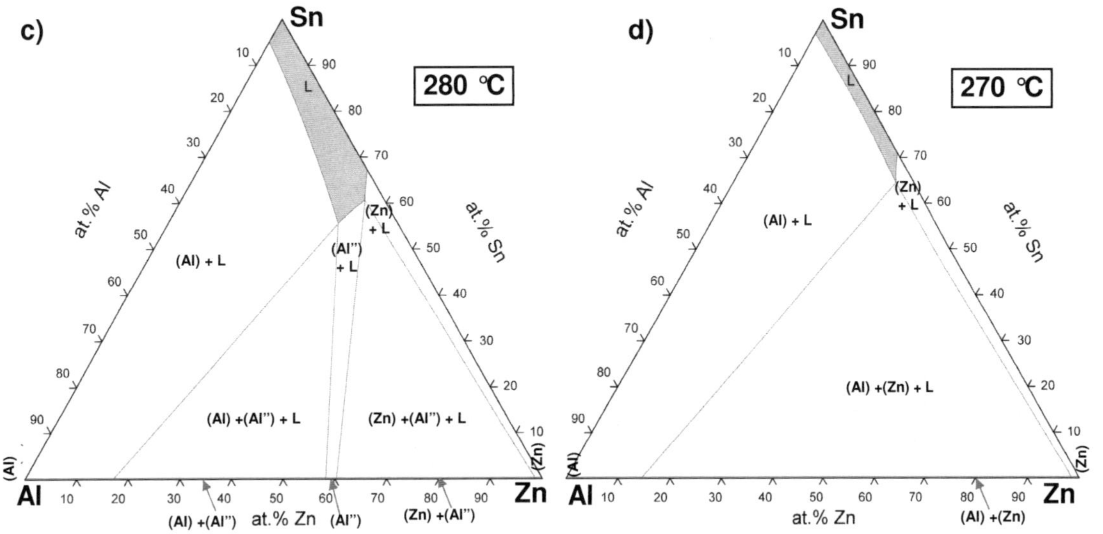

Figure 2. Isothermal sections in Al–Sn–Zn ternary system at temperatures 300 °C by [7, 10] and at 290 °C, 280 °C a 270 °C by [11].

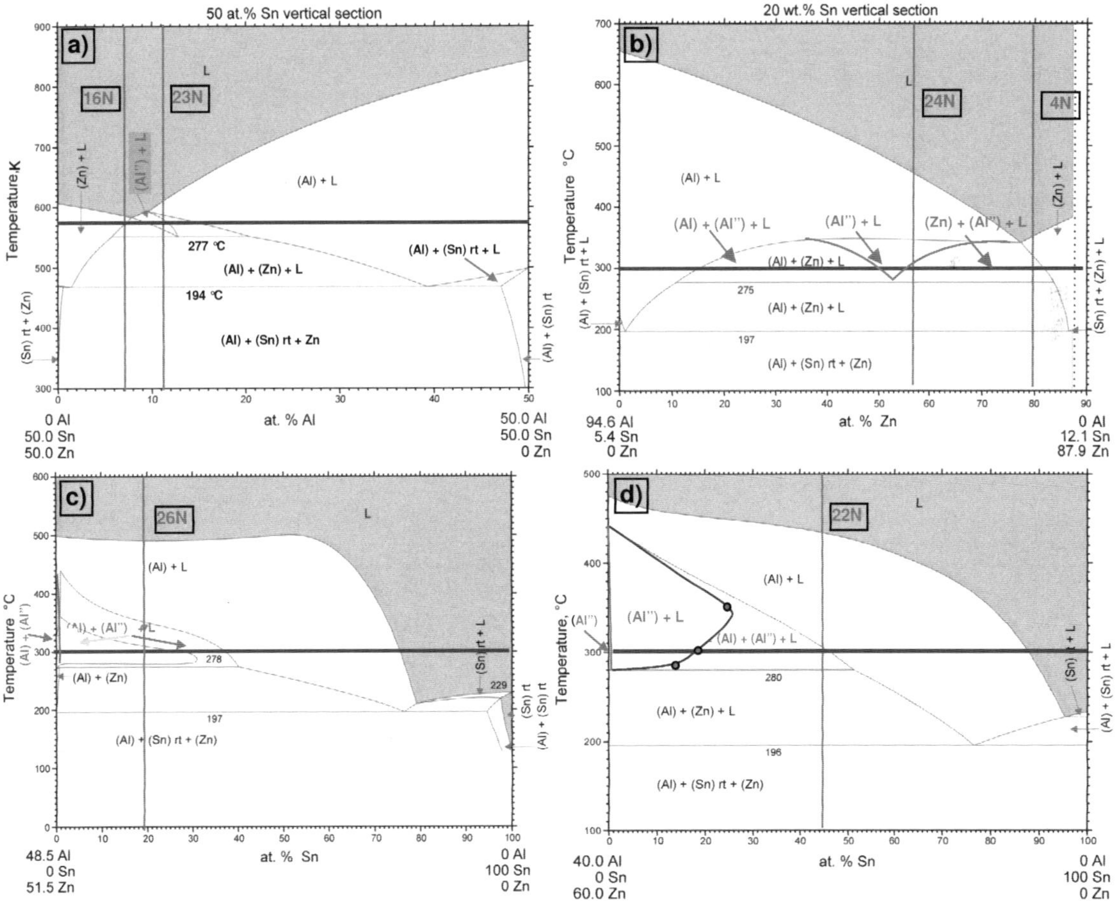

Figure 3. Vertical sections in Al–Sn–Zn ternary system a) red line „a" in Figure 1 [7, 10]; b) green line„b" in Figure 1 [3]; c) light blue line „c" in Figure 1 [3]; d) orange line „f" in Figure 1 [11].

Figure 3 depicts 4 different isoplethic sections in the ternary Al–Sn–Zn system which has been marked out in Figure 1. Figure 3 a) shows a cross-section for 50 at. % Sn according to authors [7, 10]. On confronting this diagram with isothermal cross-sections in Figure 2, we may state that the given vertical cross-section corresponds to reality. In the diagram in Figure 3 b) Prowans [3] has marked out two areas with a congruent designation (Al) + (Zn) + L, though a phase transformation occurs at the temperature of 275 °C. However, this contradicts the Gibbs phase rule as well as laws of thermodynamics. Therefore we added two thermodynamically conforming curves into the diagram in the temperature area of 275 to 350 °C. The succession of phase formation in Figure 3 b) left to right is as follows:(Al) + L → (Al) + (Al") + L → (Al") + L→ (Zn) + (Al") + L → (Zn) + L. The isopleth section in Figure 3 b) conforms to the Sebaoun ideas – see Figure 2 b), however, considering the diagrams in Figures 2 a), c) and d), the tin solubility in the area of (Al) + (Al") phases is negligible, so there is a disproportion here again. In Figure 3 b), the authors' scheme is plotted as the red curve and (Al) + (Al") + L area includes both the existing yellow tinged area above the temperature of 278 °C and the white area (marked with a green arrow), which has been considered an area of the coexistence of (Al) + (Al") phases according to [3, 11]. In Figure 3 d) all the existing phases are not complemented either. For example for the temperature of 300 °C, the succession of phase formation left to right is as follows: (Al") → (Al") + L → (Al) + (Al") + L→ (Al) + L → L. In Figure 3 b), the authors have complemented the missing curves for the area of the coexistence of (Al") phase – the red curve and for (Al") + L area – the violet curve. These complemented isopleth sections have a considerable significance for understanding the succession of phase formation during crystallization of particular alloys.

3. Experiment

3.1. Experimental samples

Six binary Sn-Zn and fourteen ternary Al-Sn-Zn alloys with 0.6 to 3 wt.% Al were prepared experimentally (melting in the electrical resistance furnace and casting into graphite mould). The purity of the initial metals was 3N5 at minimum. Next alloys with different ratios of individual elements Al, Sn and Zn (marked with the letter N – see Table 3) were prepared experimentally by smelting of pure metals in an electrical resistance furnace in evacuated quartz ampoules. These Al–Sn–Zn alloys were melted at the temperature of 750 °C in a muffle furnace, held on this temperature for 15 minutes, overturned for homogenization and held on the temperature for another hour. Finally were the samples overturned again and left in the furnace to cool down until the temperature of 70 °C was reached [12 - 15].

The selected specimens were also subjected to long-term annealing in evacuated ampoules at temperatures of 250, 300 and 350 °C for periods of 3, 7, 14 and 28 days followed by water quenching. The alloys were studied metallographically after casting and after homogenization annealing, their micro-hardness was measured, their overall chemical composition was analysed (ICP-AES and OES on the SPECTROMAX device) and X-ray micro-analysis (EDX, WDX) of the coexisting phases was performed, too.

3.2. Differential thermal analysis

The differential thermal analysis (DTA) of all the specimens was performed on the SETARAM SYSTEM 18_{TM} apparatus. The analyses were carried out twice at the speed of heating as well as cooling 4 K/min. Table 3 presents the final results of the DTA measurements of liquid temperatures T_L and eutectic transformation T_E of all the specimens including the determination of phase transformation enthalpies. The eutectic reaction temperature in binary alloys was 199 °C, while in all the ternary alloys it was $T_E = 197.7 \pm 0.7$ °C. Another invariant reaction was found at the temperature of $T_U = 278.6 \pm 0.7$ °C in some specimens. It was caused by the aluminum presence in the area of medium and high Zn concentrations. Further transformations T_1 and T_2 occur in ternary alloys between the temperatures 278 °C and liquid, which requires a more detailed analysis utilizing further methods.

The experimentally obtained transition temperatures are presented in the calculated isopleth for 3 wt.% of Al in Figure 4. Excellent agreement was achieved in the case of ternary invariant reactions temperatures [16-18].

Table 3. Nominal chemical composition of Al–Sn–Zn alloys (charge) and results of DTA analysis (heating rate 4 °C/min).

Specimen	Sn (at.%)	Zn (at.%)	Al (at.%)	T_E (°C)	T_U (°C)	T_1 (°C)	T_2 (°C)	T_L (°C)
1	5.5	87.2	7.3	197.1	279.3		360.1	385.0
2	18.2	73.8	8	197.8	279.0		346.8	355.5
3	33.7	57.4	8.9	198.9	278.3		337.3	342.3
4	52.9	37.1	10	198.8	278.7	294.1		324.0
5	70.8	18.2	11	198.6		241.0		360.0
6	5.8	94.2	0	199.1				397.0
7	19.1	80.9	0	199.4				382.0
8	35.5	64.5	0	198.9				357.3
9	56.2	43.8	0	198.8				323.5
10	75.7	24.3	0	198.9				272.6
11	82.5	15	2.5	196.6		207.1		263.0
12	60	33	7	197.4	279.8	294.2		342.4
13	70.5	24.8	4.7	197.5		258.1		324.3
14	10	80	10	197.2	278.1		355.7	363.1
15	39	51.8	9.2	197.8	278.7		324.0	348.5
16	50	43	7	197.6	279.1		310.2	333.0
17	91	6.5	2.5	196.9		215.1		268.0
18	85	15	0	198.0				
19	80	12	8	198.4		207.8		379.4
20	25	65	10	198.1	278.2		341.4	346.0
1N	85	12.5	2.5	198.8				
3N	70.4	24.7	4.9	197.6				253.8
4N	10	80	10	196.7	277.3	327.1		358.8
10N	70	20	10	197.6				239.0
12N	60	33	7	197.9				296.5
15N	39	52	9	197.3	277.5			325.0
16N	50	43	7	197.9	277.5			313.6
21N	20	65	15	197.5	277.6	325.3	363.4	439.6
22N	40	35	25	197.5	277.8	297.0	318.6	406.0
23N	40	42	18	197.6	277.8	311.6		337.0
24N	9.8	57.9	32.3	196.9	277.4	344.5	~412	446.9
25N	18.7	29.8	51.5	197.4	277.4	333.8	~439	536.4
26N	20.5	39	40.5	197.8	277.4	302.9	~332	506.5

IOP Conf. Series: Materials Science and Engineering **266** (2017) 012002 doi:10.1088/1757-899X/266/1/012002

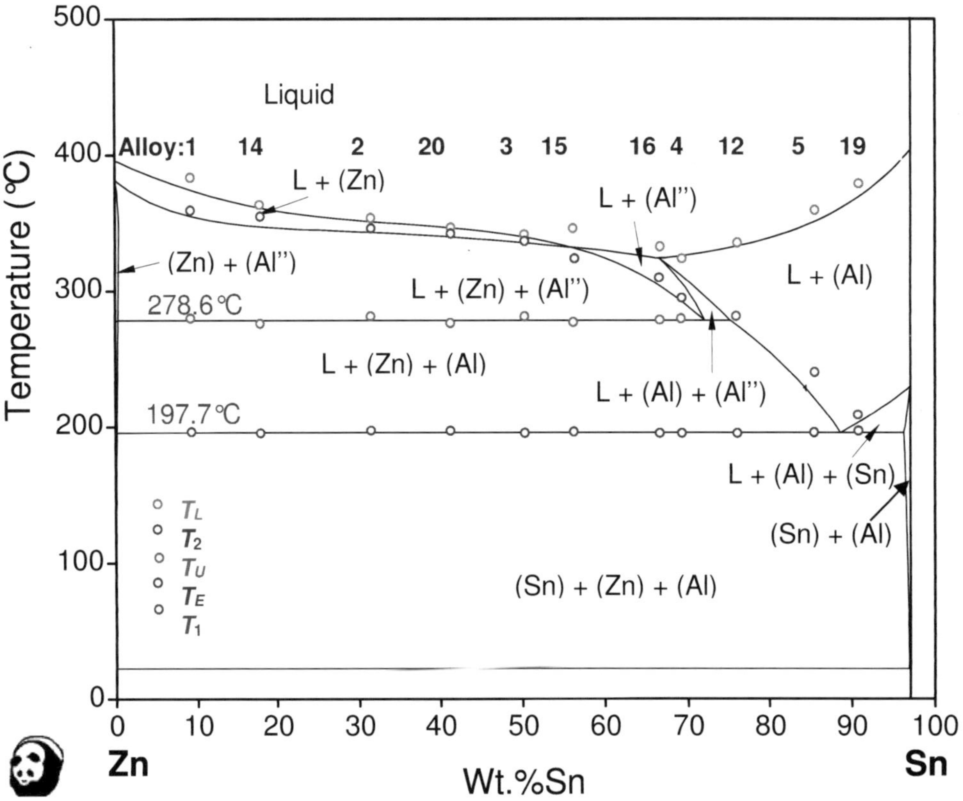

Figure 4. Vertical section in Al–Sn–Zn ternary diagram for 3 wt.% Al with the experimental values of temperatures (DTA analysis).

3.3. Microhardness

The Vickers microhardness test HV was performed. The results of HV 0.1 measurements (load 100 g) are plotted in Figure 5 in dependence on the tin content for both the binary Sn–Zn system and ternary Al–Sn–Zn alloys containing 3 wt.% Al. It can be concluded from measurements that the microhardness decreased with the rising tin content in all hypoeutectic alloys. However, aluminum increased microhardness substantially, especially with high zinc contents, when the microhardness value was nearly double compared to the binary tin-zinc system.

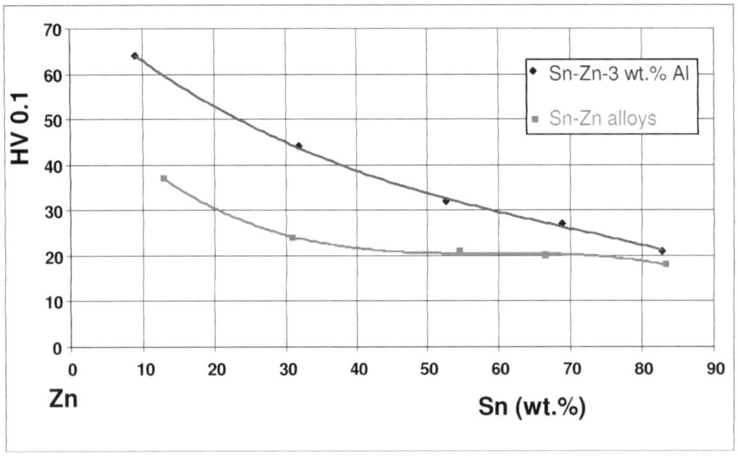

Figure 5. Dependence of the microhardness HV 0.1 on the tin content in binary Sn–Zn system and ternary Al–Sn–Zn alloys containing 3 wt.% Al.

IOP Conf. Series: Materials Science and Engineering **266** (2017) 012002 doi:10.1088/1757-899X/266/1/012002

3.4. Metallographic study

The metallographic analyses showed basically fine-grained structure, mostly of a eutectic type, often with present dendritic formations. A certain portion of fine lamellar ternary eutectics (Al)+(Sn)+(Zn) with a minority Al presence was identified in all the samples. Needles of primarily precipitated Zn and the secondary Al–Zn alloy containing 9÷12 at.% Zn were found in the structures – see Figures 6 and 7.

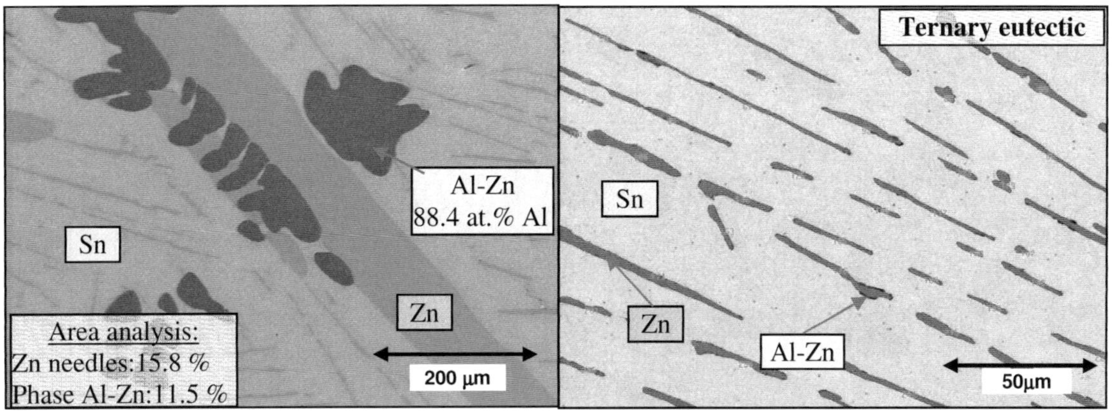

Figure 6. SEI/EDX structural and phase analysis of the sample 16N (7 at.% Al, 43 at.% Zn, 50 at.% Sn), liquid temperature T_L = 314 °C.

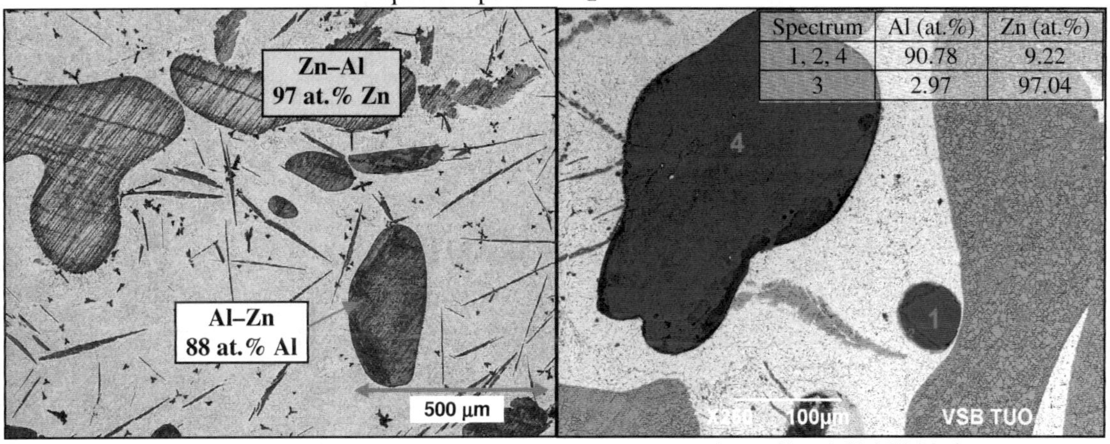

Annealing 250 °C/3 days Annealing 250 °C/ 28 days

Figure 7. Microstructural and phase analysis of the sample 16N after annealing at 250 °C in evacuated ampoule followed quenching into water.

In Figure 6 on the lefts shown a microstructure of the alloy 16N (charge: 7 at.% Al, 43 at.% Zn, 50 at.% Sn) after casting – initial state. In Figure 6 on the right is presented a microstructure with eutectic composition. Main part was formed by pure tin, on the thin lamellas of pure zinc were sometimes seen particles of (Al) solid solution. Area fraction of the Zn + (Al-Zn) phase in eutectic was 9.6 %. Microhardness of the eutectic was $HV_{0.01}$ = 15. In Figure 7 is shown microstructure of the alloy 16N after long-time heat treatmentat the temperature 250 °C – 3 and 28 days. It can see a matrix with eutectic and two phases (Al) with 10 – 12 at.% Zn and (Zn) with 3 at.% Al. Presence of tin in these phases has not been observed. In Figure 8 is shown a microstructure of the alloy 15N (charge: 9 at.% Al, 52 at.% Zn, 39 at.% Sn) after annealing at the temperatures 300 °C/14 days and 250 °C/28 days. In the microstructure were found: ternary eutectic with high content of tin, then solid solution (Zn) with aluminum (3 and 6 at. %). At temperature 250 °C was in structure present a phase (Al) with 11 at.% Zn and at 300 °C was found a phase (Al") with 62.3 at.% Zn.

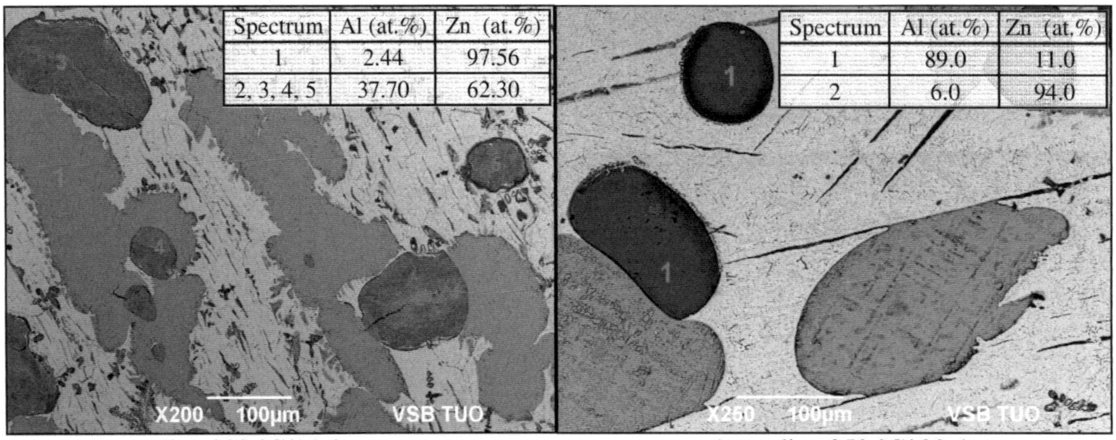

Spectrum	Al (at.%)	Zn (at.%)
1	2.44	97.56
2, 3, 4, 5	37.70	62.30

Spectrum	Al (at.%)	Zn (at.%)
1	89.0	11.0
2	6.0	94.0

Annealing 300 °C/14 days Annealing 250 °C/ 28 days

Figure 8. SEI/EDX structural and phase analysis of the sample 16N (7 at.% Al, 43 at.% Zn, 50 at.% Sn), liquid temperature T_L = 314 °C.

3.5. Isothermal diffusion in ternary system Al–Sn–Zn

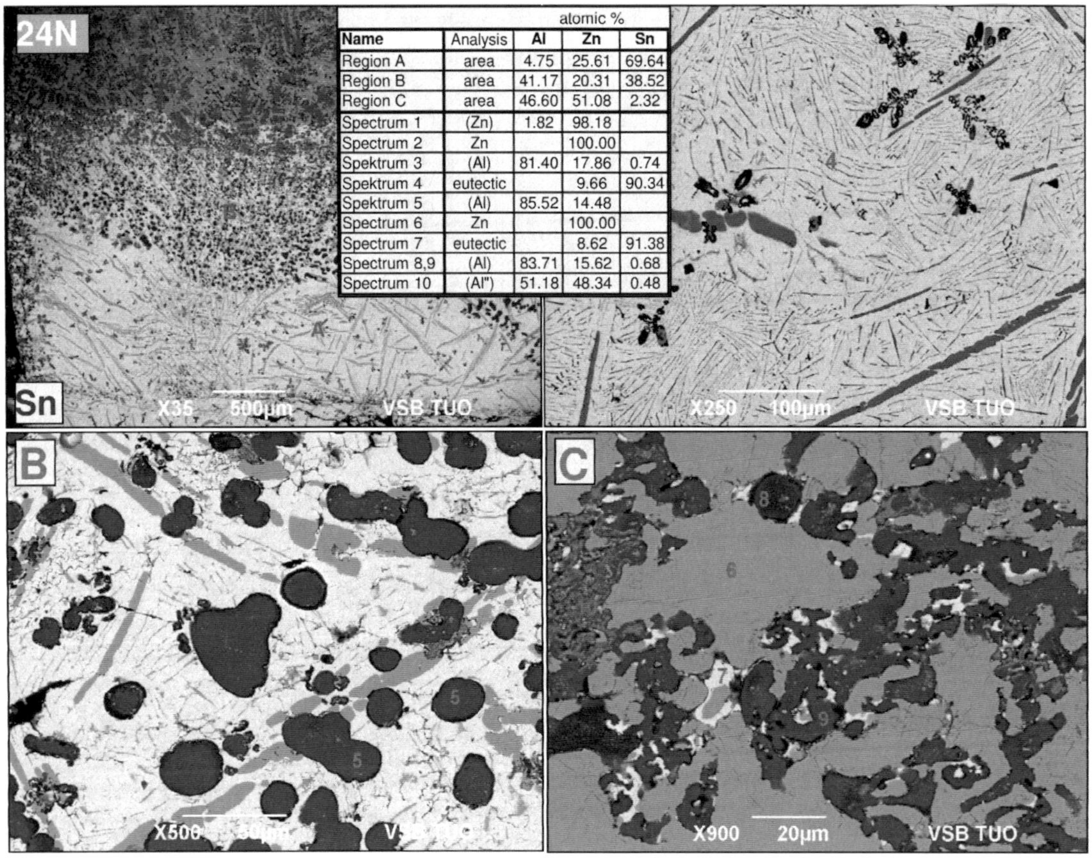

		atomic %		
Name	Analysis	Al	Zn	Sn
Region A	area	4.75	25.61	69.64
Region B	area	41.17	20.31	38.52
Region C	area	46.60	51.08	2.32
Spectrum 1	(Zn)	1.82	98.18	
Spectrum 2	Zn		100.00	
Spektrum 3	(Al)	81.40	17.86	0.74
Spektrum 4	eutectic		9.66	90.34
Spektrum 5	(Al)	85.52	14.48	
Spectrum 6	Zn		100.00	
Spectrum 7	eutectic		8.62	91.38
Spectrum 8,9	(Al)	83.71	15.62	0.68
Spectrum 10	(Al")	51.18	48.34	0.48

Figure 9. SEM/EDX microanalysis of the diffusion couple: pure tin-alloy 24N (32 at.% Al, 58 at.% Zn, 10 at.% Sn) after vacuum annealing 300 °C/6 hours followed by quenching into water.

Phases formed by isothermal diffusion in the Al–Zn–Sn ternary system have been studied at 300 °C for the three various couples (Al–Zn–Sn)-Sn. The experiment completes a study using direct together with differential thermal analysis, and provides better understanding of isothermal equilibria. Comparing the results allows different possible diffusion paths to be proposed, and EDX microprobe analysis of phases in isothermal equilibrium gives detailed information on the Al–Zn–Sn phase diagram. As an example we show a profile of the diffusion couple: pure tin-alloy 24N (32 at.% Al, 58 at.% Zn, 10 at.% Sn) after vacuum annealing 300 °C/6 hours followed by quenching into water – see Figure 9. Region A was before the heat treatment a pure tin. During annealing diffused here Zn and Al from alloy 24N. We found in the structure: eutectics, needles of pure Zn and phase (Al) with c. 18 at.% Zn. Transition region B contained: ternary eutectic, pure Zn and phase (Al) with c. 15 at.% Zn. Region C with high concentration of Zn contained minority portion of eutectic and two phases: (Al) with c. 15 at.% Zn and (Al”) 48 at.% Zn. It means that in region C were at the temperature 300 °C in equilibrium three phases: (Zn) + (Al) + (Al”).

4. Thermodynamic modeling of Al–Zn–Sn ternary system

Theoretically calculated isothermal sections (software MTDATA) at 300 and 250 °C in the Al–Sn–Zn ternary system are presented in Figure 10. The positions of the chemical compositions of some alloys are noted in these diagrams together with the results of area structural analyses of individual phases. We found in all the studied alloys the presence of the ternary eutectic (Sn) + (Zn) + (Al) with high concentration of tin. Then long needles of pure zinc were found in the microstructure of alloys. Next phases were: solid solutions (Al) with content about 8-12 at.% Zn and in some cases a solid solutions (Al”) with 50-60 at.% Zn.

On the basis of the DTA, structural and EDX analyses of Al–Sn–Zn alloys after crystallization and after long-time annealing in evacuated ampoules (250, 300 and 350 °C; times 3, 7, 14 and 28 days) followed by quenching into water were observed the proportions of the individual phases in the structure. The results of the phase equilibria for annealing temperatures 300 and 250 °C are presented in Figure 10 (examples for alloys 4N, 15N and 16N). The positions of the chemical compositions of some alloys are noted in these diagrams on the basis of results of area structural analyses of individual phases [16, 17].

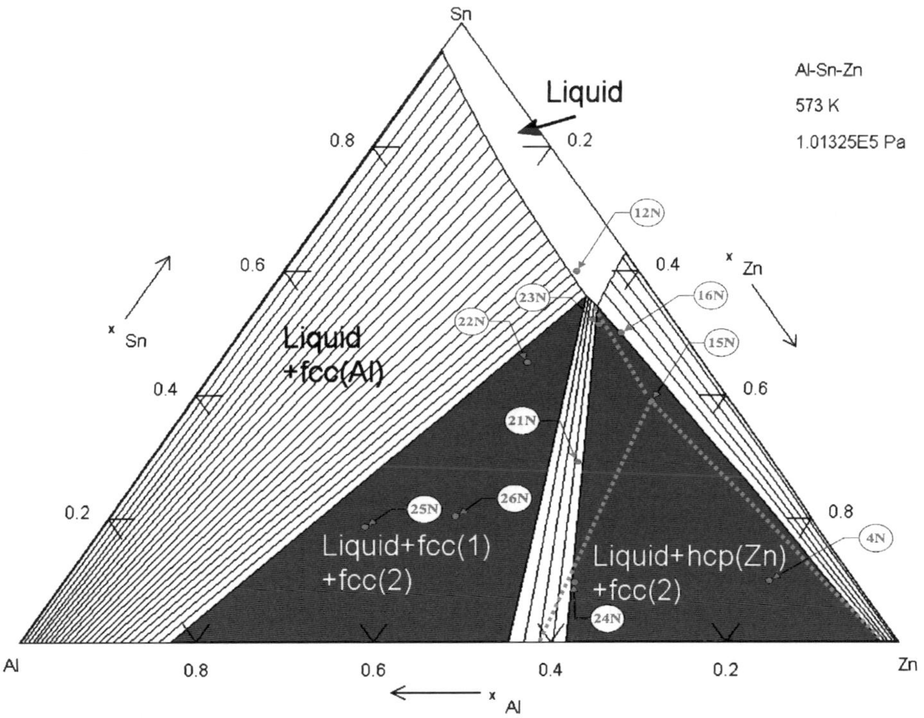

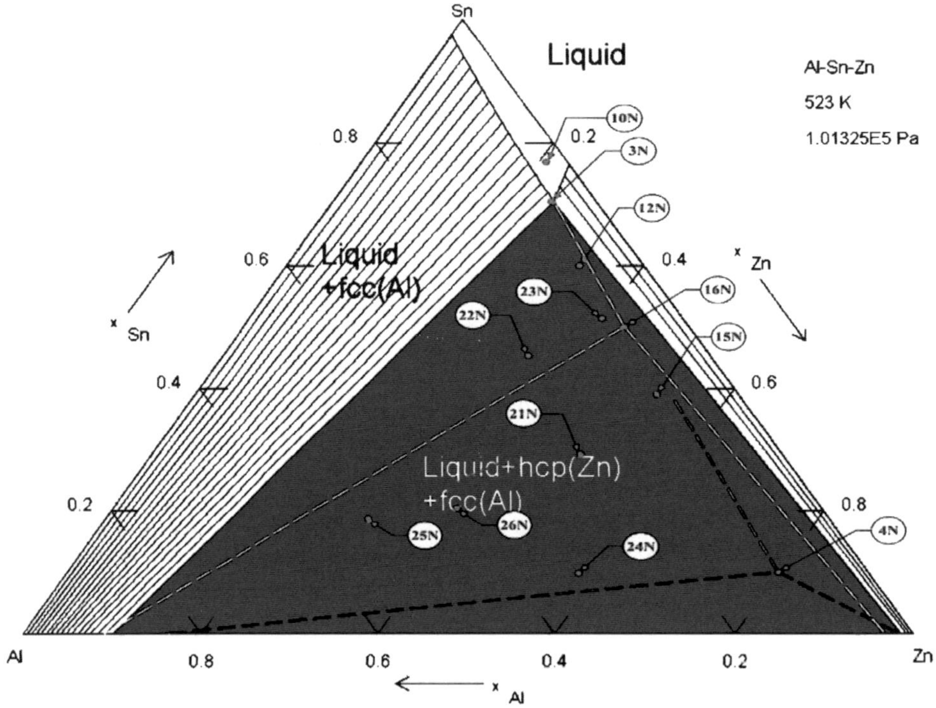

Figure 10. Isothermal sections in Al–Sn–Zn ternary diagram for temperatures 300 and 250 °C with the positions of experimental alloys and phase equilibrium (area analyses of structures after long-time annealing). Notes: fcc(Al) = fcc(1) is solid solution (Al), fcc(2) is solid solution (Al”), hcp(Zn) is solid solution (Zn).

5. Physical and technological properties of Al–Sn–Zn alloys

5.1.Electrical resistivity

Resistivity of the selected specimens of Al–Sn–Zn alloys was measured by two methods: the contact method (two-point and four-point) in accordance with DIN/IEC 468 [19] and the contactless method for measuring in a rotating magnetic field. The specimens were adapted to a unified shape in order to ensure reproducibility of the measurements. The results of the contact method measurements are shown in Table 4. It was found out that resistivity in the Al–Sn–Zn alloys was lower than in **P5** solder of SAC type commonly used today.

Resistivity of metals depends on purity of particular elements and on structural indefectibility of materials. It can be changed significantly even by a negligible amount of impurities. Comparing the results of our own measurements of resistivity with literary data, for example for Sn–Pb alloy $\rho = 14.4$ μΩ·cm according to [20] and 15.0 μΩ·cm according to [21], or 14.27 μΩ·cm [22] compared to 15.3 μΩ·cm resistivity measured in our experiment (see Table 4).

Table 4. Electrical resistivity (μΩ·cm) of solders

Alloy	Composition (wt.%)	Resitivity ChU	Resistivity CTU
P5	Sn95.5 Ag3.8 Cu0.7	11.98	11.7
P6	Sn63 Pb37	15.26	16.2
5	Sn84 Al2.7 Zn13.3		10.8
15	Sn56 Al3 Zn41		9.34

Notes: ChU – measured at Charles University in Prague, CTU – Czech Technical University in Prague

Table 5. List of alloys for measurements.

Sample no.	**Zn** (at.%)	**Sn** (at.%)	**Al** (at.%)	T_L (°C)
17	6.5	91	2.5	268
19	12	80	8	379
13	25	70	5	324
16	43	50	7	333
20	65	25	10	346

IOP Conf. Series: Materials Science and Engineering **266** (2017) 012002 doi:10.1088/1757-899X/266/1/012002

From this follows that the values measured in our experiment were by 12 % higher than literary data in average. This may be caused by an initial condition of our specimens which were deformed by cold rolling to foils of 0.2 mm thickness and were not annealed for internal defects reduction before the measurements. Nevertheless, the results can be considered objective and comparable to one another because all the specimens had the same geometric shape, dimensions and processing history.
The resistivity temperature dependences (see Figure 11) were linear in all the Al–Sn–Zn alloy specimens and were fitted with a linear function: $\rho = \rho_T[1+\beta(T-T_L)]$, where ρ_T stands for the molten metal resistivity at T_L temperature, β stands for the temperature coefficient of specific resistance.
Temperature dependences $\rho(T)$ were also linear for the solid phase and were fitted with a similar linear function. $\rho(T)$ curves almost merged with each other during heating as well as cooling in the liquid state (molten metal). Certain hysteresis was observed in some specimens during the change from the solid to liquid state and vice versa. The total uncertainty for determination of resistivity values was determined ± 5 % [23, 24].

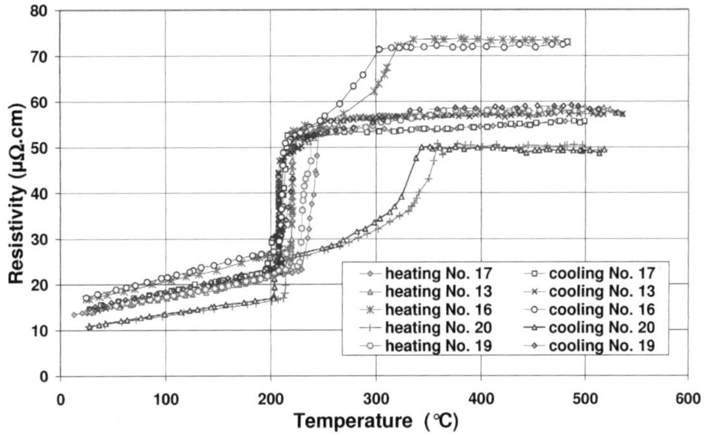

Figure 11. Temperature dependence of the electrical resistivity of some Al–Sn–Zn alloys.

5.2.Density
Figure 12 shows the temperature dependence of density for five Sn–Zn–Al alloys. Alloy no. **17** (high Sn content) exhibits the highest density. It is not without interest that a pair of alloys no. **19** and **13** and alloys no. **16** and **20** have approximately the same density at different zinc contents (see Table 5).

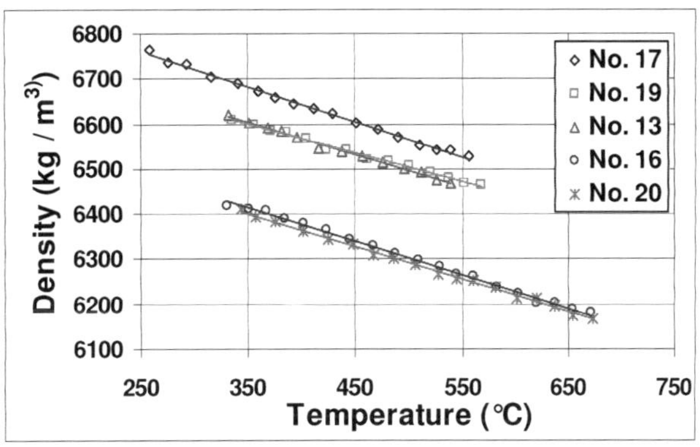

Figure 12. Temperature dependence of density of some Al–Sn–Zn alloys in liquid state.

5.3. Magnetic susceptibility

Magnetic susceptibility χ was measured using the Faraday method during heating and subsequent cooling at a rate of 3 K/min in a helium atmosphere in BeO crucibles [23, 24]. Pressure in the vacuum chamber was 10^{-3} Pa. In all the measurements it was found out that the Al–Sn–Zn alloys featured only low susceptibility to the outer magnetic field, therefore the uncertainty in susceptibility absolute values was on ± 10 % level. For all the specimens $\chi > 0$ up to $T = 200$ to 210 °C (the solid phase area) and above this temperature $\chi < 0$ (for specimen no. **20** $\chi < 0$ at $T > 140$ °C, however, a decrease was also observed at 210 °C). Other anomalies were not found at liquid temperatures, however, worth noting are temperature regions $T = 380$ to 460 °C and 530, or 580 °C, where susceptibility was constant, which was an evidence of certain structural rearrangements. The graph (see Figure 13) shows here evidently a step change of χ during the transition from the solid state to liquid.

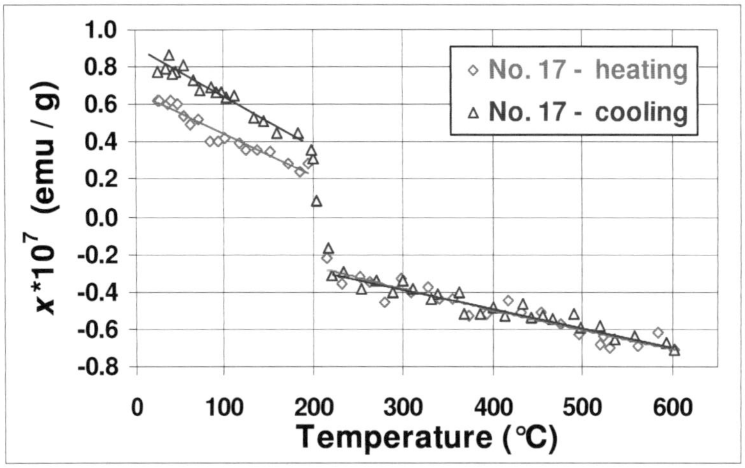

Figure 13. Temperature dependence of the magnetic susceptibility of Al–Sn–Zn alloy no. **17**.

5.4. Wettability

Wettability is defined as an ability of a molten solder to adhere to a surface of the material being joined in the presence of a flux at a specified temperature. A formation of an intermetallic compound is an essential condition of good wettability and the bond between a solder and a wetted metal.

The wetting balance method was used, enabling to obtain accurate quantitative results and observing also a non-equilibrium situation. A tested component was suspended on a dynamometer above a pot filled with a molten solder. The measurement of wettability was carried out on Meniscograph Solderability Tester GEC in CTU in Prague. The apparatus consists of a measuring head with a spring microbalance, a holder for clamping the tested object, a solder bath and an electronic control unit [25].

Regarding relatively small amounts of solder specimens, the measuring apparatus was modified in the part of the molten bath, so that the alloy amount needed to perform the test was reduced to approximately 8 cm^3. Crucibles, made of stainless steel and graphite, were used for this purpose. A small crucible with a specimen of the Al–Sn–Zn alloy was lowered and dipped into the main Pb–Sn solder bath (the eutectic composition) of a volume of c. 250 cm^3 in the tester pot and clamped in holders made of stainless material laid over the original pot.

Eight specimens were selected for the Al–Sn–Zn alloys wettability testing. Three types of fluxes were used for testing of wettability of solders with various substrates on a meniscograph. Almost for all the tested combinations: a solder, a material, a flux – a perfect wetting of a specimen by a solder has not occur. Non-wetting of the specimens of Cu and Ni wires resulted from the used type of fluxes, which appeared insufficient in term of the used measurement temperatures of molten solders. An example of a non-wettable course is depicted in Figure 14.

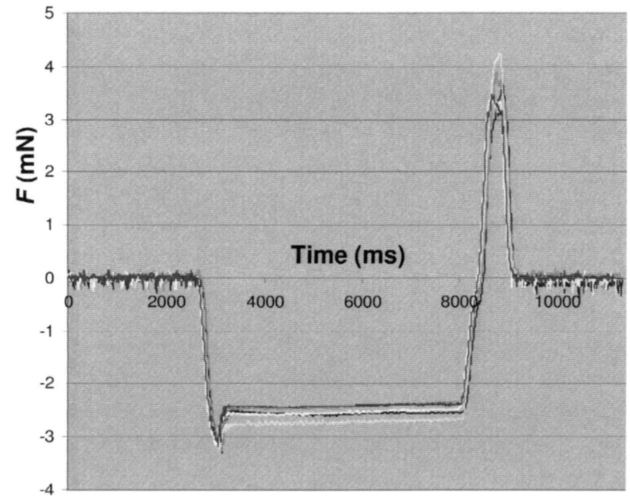

Figure 14. Wetting forces of the alloy 17.8 wt.% Sn, 78.2 wt.% Zn, 4 wt.% Al; Ni wire ø 1 mm; flux RXZ (94-RXZ-M – flux on the basis of synthetic resin formachine soldering).

6. Discussion

When testing resistivity of solders, it was found out that the Sn–Zn–Al alloys had ρ values lower than Pb–Sn solder and even lower than the best known alternative solder SAC, whereas the presence of aluminum supported the resistivity decrease. Our prepared specimens were deformed by rolling before the testing, which resulted in an increase of ρ by approximately 12 % in comparison with the non-deformed condition. A linear increase in resistivity in the solid as well as in the liquid state (molten metal) was observed from temperature dependences $\rho(T)$ in five investigated alloys – see Figure 11. According to expectations, a resistivity step increase occurred in all the solders in the area of the eutectic reaction temperature. Complicated reactions at phase transformations occurred in specimens of alloys no. **16** and **20** with high zinc contents (Table 5). A certain hysteresis degree was evident on the curves.The lowest ρ value was exhibited by alloy no. **20** with a high Al content (10 at.%).

A linear decrease $\rho_L(T)$ was observed from temperature dependences of molten metal densities in the above mentioned five specimens – see Figure 11 which was in conformity with theoretical assumptions. An interesting finding was that alloys no. **19** and **13** and alloys **16** and **20** nearly overlaid one another, which was caused by the presence of aluminum in a suitable ratio to Sn and Zn.

The magnetic susceptibility measurement confirmed that this method was sensitive to structural and phase transformations, too. Here also, a jump of χ values at the eutectic temperature and a certain degree of hysteresis behavior during heating and cooling of the specimens became evident (Figure 13).

Three types of fluxes were used for testing of wettability of solders with various substrates on a meniscograph. Cu and Ni wires (1 mm diameter) were applied as substrates and 8 Sn–Zn–Al alloys were tested. The measurements proved that nickel exhibited higher wettability with the solders used, however, the measurements results showed low wettability in general compared to common Pb–Sn, SAC and Sn-Cu solders. Here the presence of aluminum in the alloy had a negative effect; aluminum has a tendency to form oxides on solder surfaces and their dissolution using the particular fluxes failed. The results of measured courses of wetting forces "on plateau" (Figure 14) imply that the Sn–Zn binary alloys can be preferred to aluminum containing alloys. These conclusions were confirmed also by the results of a subjective image analysis. It can be stated that solders not containing aluminum are better solderable. The solderability decreases with an increasing proportion of aluminum in a soldering alloy. An issue remains to find an applicable flux even for these types of alloys. In this case, it will be advisable to use nitrogen as the inert atmosphere for testing.

7. Conclusion

The presented paper deals with a complex study of the Al–Sn–Zn ternary system both in term of phase equilibria and structural and mechanical characteristics depending on the alloy chemical composition and a method of heat treatment (a condition after crystallization or long-term annealing in vacuum at three different temperatures and time periods followed by subsequent rapid cooling – quenching). Further, measurements of resistivity, density, magnetic susceptibility and wettability of the selected alloys were carried out. Some results are presented in relevant tables and graphs. The alloys were also subjected to various corrosion tests not mentioned in this paper [26, 27].

For further applications it will be necessary to perform other technological tests of solderability to find out an optimal flux and to optimize the chemical composition of the Al–Sn–Zn based lead-free solder for particular soldered materials, especially for aluminum alloys.

Acknowledgements

This work was supported by Ministry of Education, Youth and Sport of the Czech Republic in the project EU – COST MP0602 "Advanced solders materials for high-temperature applications" No. OC 08032 and in the project No. LO1203 "Regional Materials Science and Technology Centre – Feasibility Program".

References

[1] Massalski, TD 1996 B*inary Alloy Phase Diagrams.* Second Edition Plus Updates on CD ROM (Ohio: ASM International, Metal Park)
[2] Dorward RC 1976 *Metal Trans.* **7A** 308
[3] Prowans S and Bohatyrewicz M 1968 *Arch. hutn.* **13** 217
[4] Nayak AK 1975 *Trans. Indian Inst. Met.* **28** 148
[5] Plumbridge DV 1911 *On the binary and ternary alloys of aluminum, zinc, cadmium and tin.* Thesis (Munich: University of München Germany)
[6] Losana L and Carozzi E 1923 *Gazetta Chimica Italiana* **53** 546
[7] FRIES, SG, LUKAS, HL, KUANG S, EFFENBERG, G 1991 Calculation of the Al–Zn–Sn ternary system. In "*User Aspects of Phase Diagrams*", Ed. F. Hayes, The Institute of Metals, London, 280-286
[8] http://www.msiport.com/ MSI Eureka Content - Materials Science International, GmbH.
[9] http://www1.asminternational.org/asmenterprise/apd(Ohio: ASM Alloy Phase Diagrams Center)
[10] Hubert-Protopopescu M 1993*Aluminium – Tin – Zinc* www.msiport.com
[11] Sebaoun A, Vincent D and Tréheaux D1987*Materials Science and Technology* **3** 241
[12] Drápala J et al. 2011 *Lead-free solders. Theoretical and experimental study of phase equilibria and properties of new types of solders.* Monograph (Ostrava: ES of VŠB – TU Ostrava) p 218
[13] Kostiuková G 2016 *Low-temperature alloys for lead-free solders. Study of physico-chemical characteristics.* Dissertation. (Ostrava: VŠB – Technical University of Ostrava) p. 87
[14] Drápala J, Kroupa A, Smetana B, Burkovič R, Lasek S and Musiol J 2009 Study of Zn–Sn–Al alloys for high-temperature solders. In *Metal 2009* (Ostrava:Tanger, Ltd.) 380
[15] Drápala J, Musiol J, Petlák D, Burkovič R and Vodárek V 2010 Study of behavior of tin–zinc–aluminium alloys for high–temperature applications In *Metal 2010* (Ostrava:Tanger, Ltd.) 776
[16] Drápala J, Kostiuková G, Smetana B, Madaj M and Kroupa A 2015 *Advanced Science, Engineering and Medicine* **7** 291
[17] Villain J, Klima S, Drapala J, Smetana B, Zla S, Vodarek V, Lasek S, Petlak D, Konecna K and Kostiukova G 2012 *Experimental and theoretical study of the Al-Sn-Zn system* In COST MP0602 action - Handbook of high-temperature lead-free soldering systems Group Project Reports (Brno: ed. A. Kroupa, COST Office) **3** 243-280

IOP Conf. Series: Materials Science and Engineering **266** (2017) 012002 doi:10.1088/1757-899X/266/1/012002

[18] Smetana B, Drápala J, Zlá S and Kroupa A 2009 *Proc. of 38th Int. Conf. „Calphad 2009" (Prague)* (Brno: Masaryk University and Institute of Physics of Materials AS of the Czech Republic 175
[19] *DIN/IEC Standart 768* „Method of measurement of resistivity of metallic materials"
[20] *Database for solder properties with emphasis on new lead-free solders* 2002 (Golden: National Institute of Standards and Technology and Colorado School of Mines)
[21] Cook BA, Anderson IE, Harringa JL and Terpstra RL 2002 *J. Electron. Mater.* **31** 1190
[22] Brorson G and Nylén M 1997 *Examination of enviromentally adapted tin-base alloys for replacement of tin-lead solders* 5 AFR Report (Sweden) 177
[23] Sidorov VE, Drápala J, Uporov S, Sabirzyanov A, Popel P, Kurochkin A and Grushevskij K 2011 Some physical properties of Al–Sn–Zn melts In *EPJ Web of Conferences* **15** 01022 http://www.epj-conferences.org or http://dx.doi.org/10.1051/epjconf/20111501022
[24] Drápala J, Musiol J, Petlák D, Vodárek V, Smetana B, Zlá S, Kostiuková G, Kroupa A and Sidorov VE 2011 Structural, thermodynamics and physical properties of Zn – Sn – Al alloys. In *Metal 2011*, Brno, 18.5.-20.5.2011 (Ostrava:Tanger, Ltd.) on CD ROM 1091
[25] Dušek K, Urbánek J and Drápala J 2009 *Hutnické listy* **LXII** 106
[26] Lasek S, Drápala J, Titz K and Blahetová M 2009 *Hutnické listy* **LXII** 101
[27] Lasek S, Drápala J, Blahetová M and Vrbický J 2009 *Hutnické listy* **LXII** 20

Analysis of properties laser welded RAK 40/70 steel sheets

E Evin[1], M Tomáš[1] and M Fujda[2]

[1] TU Košice, Faculty of Mechanical Engineering, Mäsiarska 74, Košice, Slovakia
[2] TU Košice, Faculty of Materials, Metallurgy and Recycling, Letná 9, Košice, Slovakia

emil.evin@tuke.sk

Abstract. Both, the ecological production and operation of vehicles demand using such materials for deformation zones' structural parts, which show some specific properties and use innovative technologies to process them. Specific requirements for functionality (strength, stiffness, deformation work, fatigue properties) are closely linked to processability (formability). In the paper are presented results for multiphase TRIP steel RAK40/70 when welded by pulse solid-state fiber laser YLS-5000. Based on microstructure analysis in the fusion zone and heat affected zone the welding parameters were optimised. The influence of laser welding on the strength and deformation properties was verified by characteristics of strength, stiffness and deformation work, as they were calculated from mechanical properties measured by tensile test and three-point bending test. The knowledge gathered in the field of laser welding influence on the strength and deformation properties of multiphase TRIP steel RAK40/70 should help designers when design the lightweight structural parts of the car body.

1. Introduction

The automotive industry appears to be a specific branch driven by social pressure to produce more and more safe, economical and environmentally friendly vehicles. From the view of costs, the most effective way to meet the requirements of EU limits for CO_2 95 g to100 km until 2021 is to reduce the car weight [1].

The largest share of the total car weight is the car body - approx. 25% of the vehicle weight. The car body is also the basic part of a car that guarantees passenger safety. One way to reduce the car body weight is modifying the car body structure and replacement the parts made of standard steel by parts made of advanced high strength steels or aluminium alloys, magnesium alloys (even their combination), composites [1]. The steel remains a dominant material for mid-size and compact class cars (B and C segment cars in EU) in the future – Figure 1. For large-size class (D, E and F segment in EU) the aluminium alloy lightweight body concept or a combination of steel and aluminium is applied. For sport cars (S segment in EU) composites are used. The car body weight reduction also offers a promising way to improve car fuel consumption. Furthermore, when the car weight is reduced about 100 kg both, the fuel consumption reduces about 0.3 l/100 km and CO_2 emission reduces about 3.6 g for mid-size cars and 0.4 l/100 km or 4.8 g CO_2 for light trucks. It should be noted, the reduction of both, fuel and energy consumption may not solve the problems of environmental and economic production, operation and disposal of cars throughout their lifecycle. Disposal of steel parts is not

Content from this work may be used under the terms of the Creative Commons Attribution 3.0 licence. Any further distribution of this work must maintain attribution to the author(s) and the title of the work, journal citation and DOI.
Published under licence by IOP Publishing Ltd

problematic. However, the disposal of composite materials and plastics is problematic from an environmental point of view [2].

Today, the car market is influenced by the car silhouette, the lines of edges, the elegance, the dynamic appearance of the body, because there are small differences in the technical equipment of the cars. In addition to design appeal, the car body role is to bear longitudinal, transverse, asymmetrical, bending and torsional loads, without visible signs, and in the event of crash, to mitigate extreme crew overloads to the lowest values and to prevent unwanted parts from entering the crew compartment [3,4]. The results of the SuperLIGHT-CAR project suggest that the application of special structural and safety elements from different types of high-strength steels (DP - Dual Phase, TRIP - Transformation Induced Plasticity, TWIP - Twinning Induced Plasticity, MART - martensitic, F-B Feritic-Bainitic, CP - Complex Phase steels, hot formed bored steels, post forming heat treated steels etc.) can improve the car body safety characteristics and reduce weight by 27% to 38% [5-10].

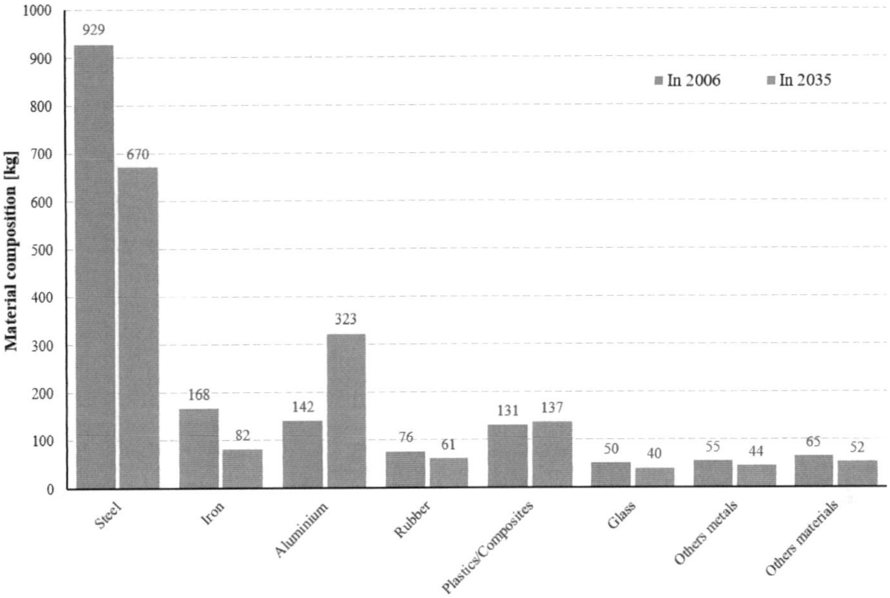

Figure 1. Materials used in automotive production and tendencies to future.

Thanks to their high energy absorption and fatigue strength, TRIP steels are widely applied in the car body deformation zones (transverse and longitudinal body beams, B-pillar reinforcements, or door sills, etc.) [11,12]. TRIP steels are multi-phase steels with transformation-induced plasticity and their microstructure consists of residual austenite islands (5-15%), bainite (25-40%) and martensite possibly scattered in the ferrite matrix (50-60%). When plastically deformed, residual austenite is transformed into martensite (TRIP effect). In order to apply the TRIP effect, a minimum residual austenitic fraction of 5-10% is required. This is ensured by increased carbon and silicon content. In recent years, increased attention has been paid to CMnAl TRIP steels, due to the increased Al content, which causes an increase in C in residual austenite. The same way as Si, Al is also insoluble in cementite, slowing its formation, and at the same time increasing the rate of bainitic transformation. The disadvantage is, Al reduces the effect of hardening of the solid solution against Si and increases the temperature of the M_S. The current development of TRIP steel production is only to partially replace with a limited amount of Al and use 0.05 - 0.10 wt% P. The amount of P used is linked to the Al content because P also suppresses cementite formation and is a very effective element in hardening of the solid solution. Higher levels of silicon and aluminum promote ferrite and bainite formation. These elements are therefore helpful in trying to maintain the required amount of carbon in residual austenite. It is important to suppress the precipitation of carbides during bainitic transformation. For this purpose, silicon and aluminum are used [13].

2. Materials and methods used

The experiments were done on multi-phase TRIP steel RAK 40/70 (Residual Austenite, K- Cold rolled, Re/Rm = 400/700). The strength, stiffness, energy absorption ability, formability and weldability have been analyzed from values of mechanical properties measured according to STN EN ISO 6892-1:2010-01; normal anisotropy ratio r measured according to STN EN 10113 and strainhardening exponent n measured according to STN EN 10275. These are shown in the Table 1.

Table 1. Mechanical properties of TRIP steel RAK 40/70 [15,16].

	Dir.	ReH [MPa]	Rm [MPa]	Ag [%]	A [%]	K [MPa]	n [-]	r [-]
	0	440	764	25.3	30.3	1497	0.295	0.66
	45	462	761	22.8	26.6	1452	0.275	0.69
BM	90	457	766	24.3	29.4	1474	0.281	0.62
	Average	453	764	24.1	28.8	1474	0.284	0.66
	Stdev	5	3	0.8	1.5	13	0.004	0.03
	0	466	657	20	21	1459	0.249	0.66
	45	483	659	19	20.1	1436	0.235	0.63
LW	90	474	662	20.3	22.3	1458	0.244	0.60
	Average	474	659	19.8	21.1	1451	0.243	0.63
	Stdev	9	2	1	1	13	0.007	0.03

BM – base material, LW – laser welded

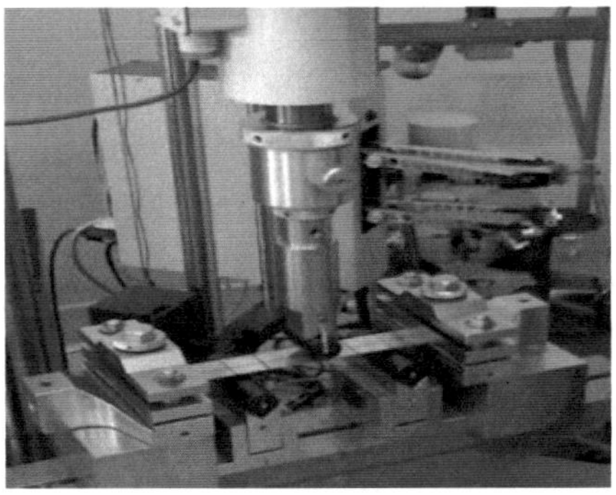
a)

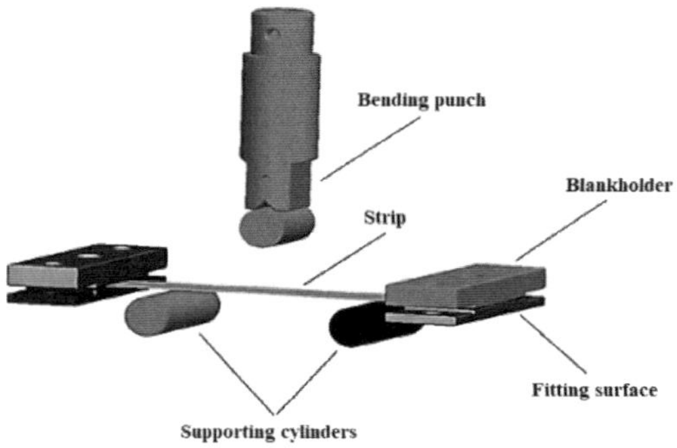

b)

Figure 2. The three point bending test: a) fixture, b) scheme.

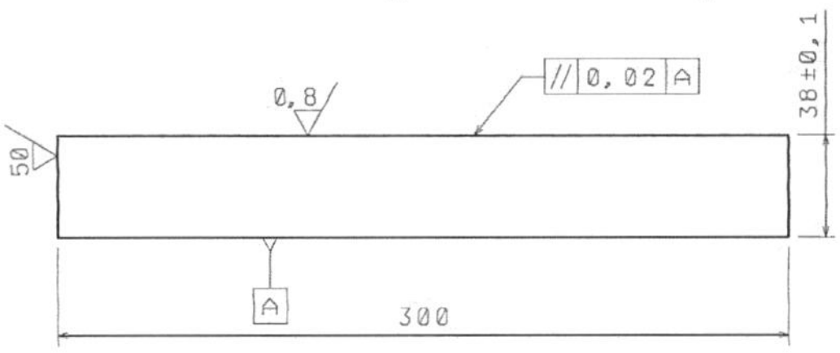

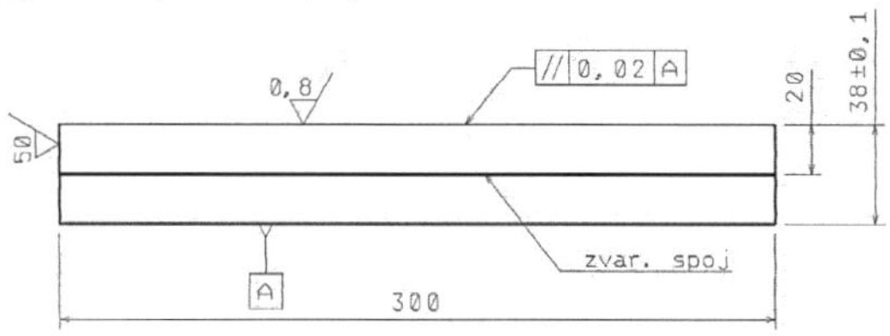

Figure 3. Samples for three point bending test – base material (left), laser welded (right).

The three-point bending test (Figure 2) was used to identify the energy absorption ability and stiffness constant c. Test were performed on samples (Figure 3) from base material (BM) and laser welded one (LW) by pulse solid-state fiber laser YLS-5000 in continuous welding regime. Parameters of laser welding are shown in Table 2 and these were optimized by metallography analysis of microstructure the weld joint (porosity, weld root quality) in the area of fusion zone and heat affected zone. The weldability of base material was evaluated by carbon equivalent calculated from equation (1). The chemical composition of TRIP steel RAK 40/70 measured by atomic emission spectrometry according to ASTM E 415-14 is shown in Table 3 [14].

Table 2. Parameters of laser welding [15,16].

Sheet thickness [mm]	Power [W]	Focal point [mm]	Welding speed [mm.s^{-1}]	Width of FZ [mm]	Width of HAZ [mm]	Note
0.7	2000	10	50	1.05	0.40	Optimal
	2700	10	70	1.17	0.48	

$$C_{ekv} = C + \frac{Mn}{6} + \frac{Cr}{5} + \frac{Ni}{15} + \frac{Mo}{4} + \frac{Cu}{13} + \frac{P}{2} + 0.0024a_0 \quad [wt\ \%] \tag{1}$$

$$C_{ekv} = 0.197 + \frac{1.576}{6} + \frac{0.0445}{5} + \frac{0.016}{15} + \frac{0.022}{4} + \frac{0.022}{13} + \frac{0.015}{2} + 0.0024x0.7 = 0.48 \quad [wt\%]$$

Table 3. Chemical composition of TRIP steel RAK 40/70 [wt%].

C	Si	Mn	P	S	Cu	Al	Cr	Mo	Ni	V	Nb	C_{ekv}
0.197	0.165	1.576	0.015	0.002	0.022	1.352	0.0455	<0.022	0.016	0.0002	0.002	0.48

3. Reached results and discussion

The weld metal microstructure consists predominantly of martensite with a morphology indicating its formation in austenite grains which are elongated in the direction of heat transfer from the center of the weld metal to the base material (Figure 4). The martensite laths in the welded metal are relatively fine and have a random orientation (Figure 4b). The fineness of the martensitic latices increases as the distance from the center line of the weld increases towards the heat-affected zone. Microhardness in the weld metal reached 470 HV0.5. A mixed martensitic-ferritic structure is present in the heat-affected zone of the welding joint (Figure 4c). Microhardness in the heat affected zone reached 450 HV0.5. The proportion of ferrite grains in this mixed microstructure increases as the distance to the base material increases. Close to the base material ferrite prevail in the microstructure of the heat-affected zone in the form of polyhedral grains and martensite in the form of fine shapes of different formation as a minor phase. The microstructure in the base material (Figure 4a) is fine grained, homogeneous with the indications of the layout of the structural components in rows. It consists predominantly of fine grains of ferrite and minority of very fine residual austenite and martensite. Structure also includes granular bainite. Microhardness in the base material reached 222 HV0.5. [15,16]

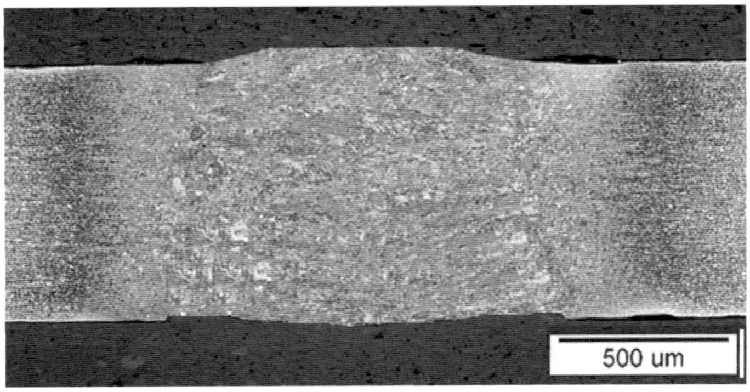

Figure 4. The microstructure of laser weld joint of steel TRIP RAK 40/70.

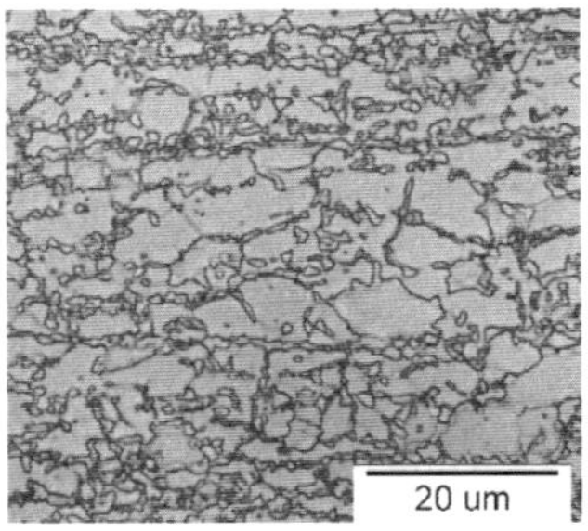

Figure 4a. Detail of the base material.

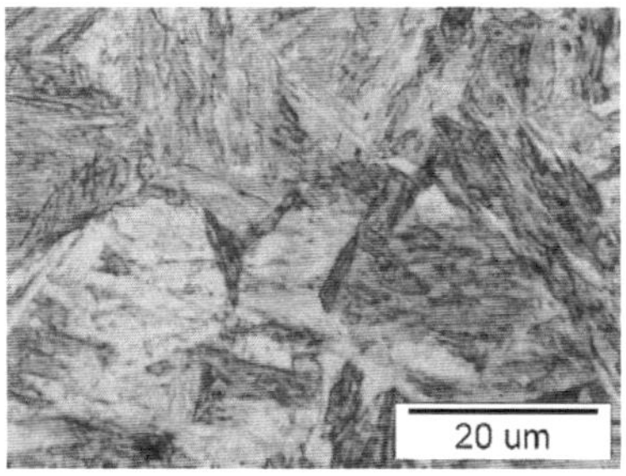

Figure 4b. Detail of the fusion zone.

Figure 4c. Detail of the heat affected zone.

The nature of the fracture area of samples broken at the tensile test and the three-point bend was similar. The edges of the fracture surfaces in the weld metal region were poorly deformed in both cases (Figure 5). On their surfaces there are dimples of transcrystalline ductile fracture and relatively smooth areas. In the heat-affected zone of both samples, the fracture surface was less rugged than in the weld metal. They formed predominantly the dimples of transcrystalline ductile fracture. The fracture of the base material was propagated perpendicular to the direction of the tensile load. Within the dimples precipitated particles or impurities were observed.

a) b) c)

Figure 5. The morphology of fracture surface a) weld metal, b) heat affected zone, c) base material

Resistance to impact for samples of base material and laser welded for TRIP steel RAK 40/70 was evaluated by total deformation work [16]

$$W_{pl} = \frac{K.S_0.L_0.(\varphi_{max,necking}-\varphi_{0.002})^{n+1}}{n+1} \qquad (2)$$

from records of testing machine (Figure 6). In the equation (2) $\varphi_{max,necking}$ means the maximal deformation (strain) of the car body component, K is the material constant, n is the strain-hardening exponent, S_0 is the sample section and L_0 is the length of sample.

The support part of the car body shall be sufficiently strong and rigid to prevent plastic deformation of the body parts as a result of forces acting in normal operation, and only small plastic deformations of the parts are permitted in the event of an accident. The strength of the base material and the laser welded samples was determined [16]

$$\sigma_{dov,t} = \frac{R_{p0.2}}{k} = \sigma_{el} \qquad (3)$$

and stiffness as the amount of elastic energy W_{el} that can absorb the material without plastic deformation of the part [16]

$$W_{el} = F.x_{\varepsilon<0.005} = \frac{R_e^2.S_0.L_0}{E.k^2} \tag{4}$$

where $R_{e0.005}$ is stress on elastic limit or $\sigma_{el} = E.\varepsilon_{el}$ at 0.005%.

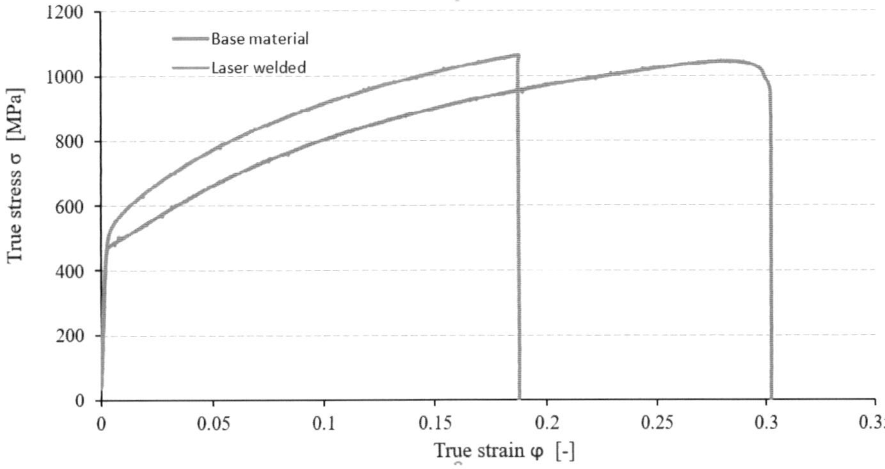

Figure 6. True stress – true strain diagram.

Similarly, it is possible to compare the stiffness and deformation work of the base material and the laser welded samples in the three-point bend test. From the record of the bending force dependence on the bend path it follows that it is a linear dependence [16]

$$F_{bend} = c.x \qquad [N] \tag{5}$$

where F_{bend} is bend force, c is the stiffness constant for defined samples shape, x is bending path or punch path. The slope of the bending force on the bend path determines the stiffness constant c – Figure 7.

Table 4. Measured and calculated values of deformation work and stiffness constant [15,16]

3-point bend test								Tensile test			
BM				LW				BM		LW	
$F_{B\,max}$ [kN]	h_{max} [mm]	W [Nm]	c [MPa]	$F_{B\,max}$ [kN]	h_{max} [mm]	W [Nm]	c [MPa]	W_{el} [Nm]	E_{pl} [Nm]	W_{el} [Nm]	E_{pl} [Nm]
13.611	38.31	296.5	0.404	22.966	35.442	424.45	0.677	0.406	190	0.444	167

In Table 4 the calculated values of W_{el}, W_{pl} and strengths according to relations (2), (3), (4) are shown. As it comes out from the results of the tensile test, laser welded samples from TRIP steel RAK 40/70 showed 9% greater stiffness (W_{el}), 5% greater strength, 12% less overall deformation work than base material samples. However, at the same deformation φ = 0.19, the laser welded samples showed greater deformation work (W_{pl} = 157/144 = 1.09) by 9%.

The results of the three-point bend test showed that the stiffness constant is 67 % higher and the total deformation work is greater by 43 % for the base material. It is supposed, the higher values of the stiffness constant and deformation work at the same deformation are given by the more intense hardening of the material in the area of the weld metal and the heat-affected zone, due to the greater proportion of martensite in the welding joint structure than in the base material.

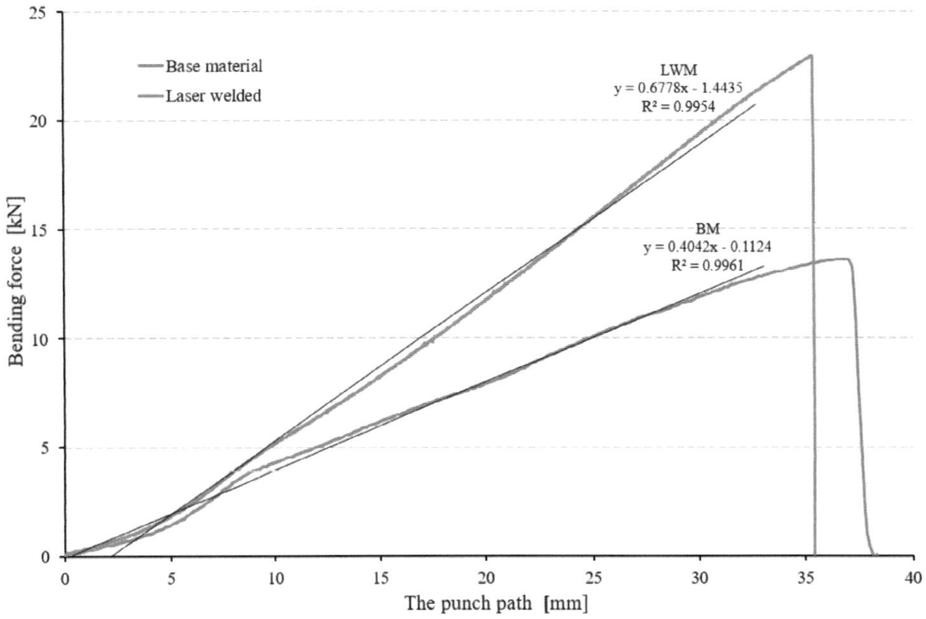

Figure 7. The bending force on the bend path dependence.

4. Conclusion

On the basis of the measured and analyzed results of the impact of laser welding by solid fiber lasers on the properties of TRIP steel sheets from RAK 40/70, it can be stated:

1. Based on the quality of the weld joint TRIP steel RAK 40/70 when laser welded by solid-state fiber laser YLS-5000 without protective gas in continuous welding mode it has been proved the most suitable mode with a welding speed of 50 $mm.s^{-1}$ and power 2000 W.
2. The microstructure of the base material multi-phase TRIP steel RAK 40/70 consists of ferrite, very fine residual austenite, martensite and granular bainite. The microstructure in the heat-affected zone is heterogeneous, consisting of martensite and ferrite. As a result of laser welding, the proportion of martensite in the form of fine shapes of different forms decreases in the ferrite matrix from the center of the welding joint to the base material. The welding microstructure of the weld metal consists predominantly of fine martensite (formation in austenite grains) with random orientation of martensite laths.
3. Welded samples of multi-phase TRIP steel RAK 40/70 showed higher yield strength, lower tensile strength, tensibility, material constant K, and strain-hardening exponent n. The decrease in strength may be due to softening in the transition zone between the heat-affected zone and the base material or the result that the entire cross-section is not homogeneous. In the area of the weld metal, the largest proportion of martensite is in the form of larger islands, and its size as well as the size of its fragments decreases towards the base metal. We assume the strength and other properties depend not only on martensite share in the mixed structure but also on its morphology and distribution in structure.

4. Strength, stiffness and deformation work are important characteristics of the deformation process. From the results of the tensile test and the three-point bend test, a slight increase in strength, stiffness and deformation performance was observed for laser welded samples. In the area of the weld metal and the heat-affected zone, a greater proportion of martensite is present in the ferrite matrix in the form of fine shapes of different forms than in the base material, which makes the strainhardening contribution in this area larger than for the base material. Laser-welded samples showed a greater deformation strength effect than samples from the base material.
5. When comparing deformation work and stiffness, the amount of 10% plastic deformation is assumed for car body parts produced by forming or deformed at crash. Thus, for laser welded samples from multi-phase TRIP steel RAK 40/70, the WH effect increases the strength by 8%, the stiffness by 17%, deformation work at tensile by 12% (73.4 / 65.6 = 1.12), and at three-point bend test the stiffness constant and deformation increase by 67%.

Acknowledgments
The authors are grateful for the support of experimental works to Slovak Research and Development Agency, under project APVV-0273-12 "Supporting innovations of autobody components from the steel sheet blanks oriented to the safety, the ecology and the car weight reduction", as well as the grant agency for the support of the project VEGA 2/0113/16 "Influence of laser welding parameters on structure and properties of welded joints of advanced steels for the automotive industry".

References

[1] Incerti I, Walker A and Purton J 2010 *Paper to the International Bodyshop Industry Symposium* (Montreux, Switzerland) p. 1-9
[2] CO_2 emmisions change the material composition of vehicles and steel share decreases [online] 2014 [citted 2015-7-22] access: <http://www.cfo.sk/articles/emisie-co2-zmenia-materialovu-skladbu-automobilov-podiel-tradicnej-ocele-klesne#.VcDyMq2r07y>
[3] Jeanneau M and Pichant P 2000 *La Revue De Metallurgie* **97** 1399-1408
[4] Jacob G C, Fellers J F, Simunovic S and Starbuck J M 2002 *Journal of Composite Materials* **36** 813-850
[5] Zhang Ch, Song X, Lu P and Hu X 2012 *Materials & Design* **36** 233-242
[6] Mihaliková M and Lišková A. 2016 *Metalurgija* **55** 753-756
[7] Singh S, Nanda T, Ravi Kumar B and Singh V 2016 *Materials and Manufacturing Processes* **31** 2064-2075
[8] Švec P, Schrek A, Hrnčiar V and Csicsó T 2015 *Acta Metallurgica Slovaca* **21** 311-320
[9] Han Z Y, Zhang M D, Xu H F, Dong H and Cao W Q 2016 *Kang T'ieh/Iron and Steel* **51** 1-9
[10] Moravec J 2016 *25th International Conference on Metallurgy and Materials* (Brno, Czech Republic) p. 303-309
[11] Chen H, Yang Y and Wang L 2015 *Open Mechanical Engineering Journal* **9** 168-172
[12] Girman V and Hrabčáková V 2012 *Materials Engineer / Materiálový inžinier* 1-10
[13] Hall J N 2011 *Great Design in Steel Seminar* (USA: Livonia, Michigan)
[14] Hlavatý I 2001 *Weldability of steels* (Ostrava: ZERROS)
[15] Evin E and Tomáš M 2017 *Metals* **7** 239
[16] Evin E and Tomáš M 2017 *Final Report of the Project APVV-0273-12* (Košice: Technical University)

Modification of Er:YbAG film microstructure with a sintering agent

V Jakeš[1], V Polák[1], K Rubešová[1], T Hlásek[1], D Mikolášová[1], L Nádherný[1], P Nekvindová[1] and J Oswald[2]

[1] University of Chemistry and Technology Prague, Department of Inorganic Chemistry, Technická 5, 166 28 Prague 6, Czech Republic
[2] Institute of Physics of the Czech Academy of Sciences, Cukrovarnická 10, 162 00 Prague 6, Czech Republic

Vit.Jakes@vscht.cz

Abstract. Thin films of erbium doped YbAG were prepared by spin-coating on fused silica substrates. The effect of two parameters on the microstructure of resulting films was observed: addition of a sintering agent (TEOS) to the solution deposited and crystallization under decreased atmosphere (100 mbar). All prepared samples were polycrystalline single phase YbAG. When comparing the reference (TEOS-free) samples crystallized in ambient pressure, the films with TEOS in a combination with decreased pressure during the crystallization had a smaller crystallite size and finer surface, as confirmed by AFM and Williamson–Hall analyses. A higher amount of TEOS caused cracking of the films though. All films containing TEOS were one-mode waveguides in the NIR region. This paper shows a way to modify microstructure of a waveguiding film via addition of a sintering agent without destroying the waveguiding ability.

1. Introduction

Garnets are widely used host materials for optical and optoelectrical applications thanks to their optical isotropy and transparency, non-toxicity, thermal and chemical stability and the possibility to tune their properties by a suitable substitution [1]. In optical applications such as lasers or active optical components, doping with rare earths (RE) cations is very often used. The luminescent properties of the RE cations can be affected by several factors, e. g. type of a host material, the concentration of the RE ions, particle size and surface morphology. Among the RE ions, Er^{3+} is one of the most often used as an optical activator. Depending on the excitation energy and Er^{3+} concentration, red [2], green [3] or infrared emission [4] of Er^{3+} can be achieved. The Er^{3+} ions have a low absorption cross-section; it is therefore desirable to use a sensitizing agent for its improvement. The photoluminescence of Er^{3+} for infrared emission is enhanced by Yb^{3+} ions as a co-dopant [5]. The Yb^{3+} cations are used as sensitizers because the energy of $F_{5/2}$ (Yb^{3+}) level is close to the $I_{11/2}$ (Er^{3+}) level. Thanks to a similar size of RE ions, the co-doping with both these ions into most of the host materials is very effective and used. Apart from garnets, other crystals with suitable optical properties are used as a host matrix, e. g. $LiNbO_3$ [6] or Y_2O_3 [7]. In YAG, all yttrium atoms can be substituted by ytterbium. When ytterbium is a part of a host matrix, every Er^{3+} ions have close neighbours for the

Content from this work may be used under the terms of the Creative Commons Attribution 3.0 licence. Any further distribution of this work must maintain attribution to the author(s) and the title of the work, journal citation and DOI.
Published under licence by IOP Publishing Ltd

energy transfer Yb^{3+} – Er^{3+}. This arrangement improves luminescent properties of Er^{3+} ions and can be used, for example, in the amplification of an optical signal in the third telecommunication window (1530 nm) [8].

Garnets and RE-doped garnets for optical applications have been produced in the form of monocrystals, powder or thin film by a whole range of methods, such as pulsed lased deposition [9], sol–gel method [10] or solvothermally [11]. The sol–gel method has proven to be a very suitable method for the preparation of thin films and powders thanks to the solution stage of the process which enables a dopant to be homogeneously distributed in a final sample. This technique has also low demands on laboratory equipment and requires relatively common chemicals such as acetates, alkoxides or nitrates [12–14]. When preparing thin films, the solution of cations is deposited on a substrate which is afterwards subjected to thermal treatment to decompose organic compounds (e. g. chelating agents) and to crystallize the film. The microstructure of thin films can be modified by optimizing the technology parameters such as temperature and pressure used to decompose and crystallize the film or by the use of a sintering agent.

To obtain a dense ceramic material with a high transparency, it is necessary to use a sintering agent and sinter the sample in vacuum. This principle is applicable also at polycrystalline films preparation. In most garnet preparations, tetraethyl orthosilicate (TEOS) is used in a small concentration $\approx$ 0.5 wt% [15, 16] as a sintering agent. TEOS can affect the sintering mechanism by lowering the sintering temperature by about 100 °C [17].

In this paper, Er:YbAG solutions were prepared with acetic acid and polyvinylpyrrolidone (PVP) used as chelating agents. The use of PVP was described by our group in the preparation of thin films of lithium niobate [18] and YbAG [19]. Thin films were deposited by spin-coating on fused silica. The effect of pressure applied during the crystallization and the influence of TEOS addition were studied.

2. Experimental

Solution to be deposited

Three solutions were prepared to deposit $(Yb_{0.98}Er_{0.02})_3Al_5O_{12}$ films with a different concentration of TEOS (0, 1, 10 wt%). Aluminium chloride, and ytterbium and erbium acetates (all supplied by Sigma–Aldrich, purity 99.9%) were used as a source of cations. The acetates were dehydrated at 200 °C prior the preparation.

$AlCl_3$ was dissolved in water to an approximate concentration of 5 g of Al^{3+} per L; the exact concentration of Al^{3+} was determined by gravimetry. To a measured volume of the Al^{3+} aqueous solution, polyvinylpyrrolidone (Sigma–Aldrich, M = 29,000 g/mol) was added in a molar ratio of 1:4 (relative to moles of Er:YbAG). After its complete dissolution, acetylacetone (Sigma–Aldrich) and acetic acid (Penta) were added in a molar ratio of 1:2:16 (relative to moles of Er:YbAG). Finally, solid erbium and ytterbium acetates were added. The entire process was performed at room temperature with magnetic stirring. After a complete dissolution of the acetates, the concentration of the solution was adjusted to 0.10 mol/L (molar concentration refers to a total concentration of metals) by evaporation at 60 °C. In the final preparation step, the sintering agent TEOS (Sigma–Aldrich) was added in two different concentrations (1 and 10 wt% with respect to the resulting Er:YbAG) and the solution was stirred magnetically for 30 min.

Deposition

Thin films of Er:YbAG were spin-coated on amorphous fused silica substrates (supplied by Crytur). The rotation speed was 3000 RPM (10 seconds) with 30 rot/s^2 acceleration/deceleration rate. The solution was deposited onto the substrate at the point of reaching the maximum rotation speed; the volume deposited was 22.5 μL per cm^2 of the substrate area. The spin-coater WS-650SZ-6NPP/LITE (Laurell Technologies Corp.) with 650 controller unit was used for the deposition of the films.

After each deposition, the layer was dried at 700 °C for 10 min. After depositing 7 layers, the film was crystallized at 1000 °C in air at ambient pressure for 1 h. The deposition process described above was proved to produce films thick enough to waveguide [19].

In order to evaluate the effect of decreased pressure at high-temperature processing, decomposition and crystallization of the second set of samples was carried out in a decreased pressure of 100 mbar (while maintaining all the remaining deposition parameters unchanged). The labelling of the prepared samples is shown in Table 1.

Table 1. Prepared samples of YbAG

Sample name	TEOS (wt%)	Pressure (mbar)
YbAG–0–1000	0	1000
YbAG–0–100	0	100
YbAG–1–1000	1	1000
YbAG–1–100	1	100
YbAG–10–1000	10	1000
YbAG–10–100	10	100

Samples characterization

The phase composition of the prepared films was determined by XRD analysis. Data were collected at room temperature with Bruker-AXS D2 Phaser powder diffractometer with parafocusing Bragg–Brentano geometry using CoK_α radiation in a range of 18–53° 2θ with a step size of 0.014°. The phase analysis was performed in the HighScore Plus program using the PDF-4 database. For the purpose of calculating crystallite size and internal stress, the linearized Wiliamson–Hall equation [20] was used.

Roughness of the prepared films was measured by AFM analysis using the Ntegra Spectra (NT-MDT) device. The surface scanning was performed in the tapping mode. Cantilevers with a strain constant of 1.5 kNm^{-1} equipped with a standard silicon tip with curvature radius lower than 10 nm was used for all measurements. The surface of the prepared films was observed using a scanning electron microscope VEGA 3 LMU (Tescan) with an EDS probe INCA 350 (Oxford Instruments) in a backscattered electron mode. The samples were coated with gold (2 nm) prior the analysis.

Waveguiding properties were measured by m-line mode spectroscopy with the Metricon Prism Coupler device. The measurement was performed using a prism with an effective range of refractive index values from 1.2 to 2.02 (for 633 nm wavelength) at the operating wavelengths of 633, 964, 1311 and 1552 nm in a TE polarization.

3. Results

As the XRD patterns in Figure 1. show, all samples were polycrystalline single phase YbAG with no secondary phases. The reflections were labelled according to the PDF-4 reference pattern No. 01-073-1369. The sample prepared at ambient pressure without the addition of TEOS (YbAG–0–1000) exhibited the lowest intensity. The measurements are presented only in a range of 2θ = (20–36)°. This range matches the range used in our earlier paper [19] for easy comparison.

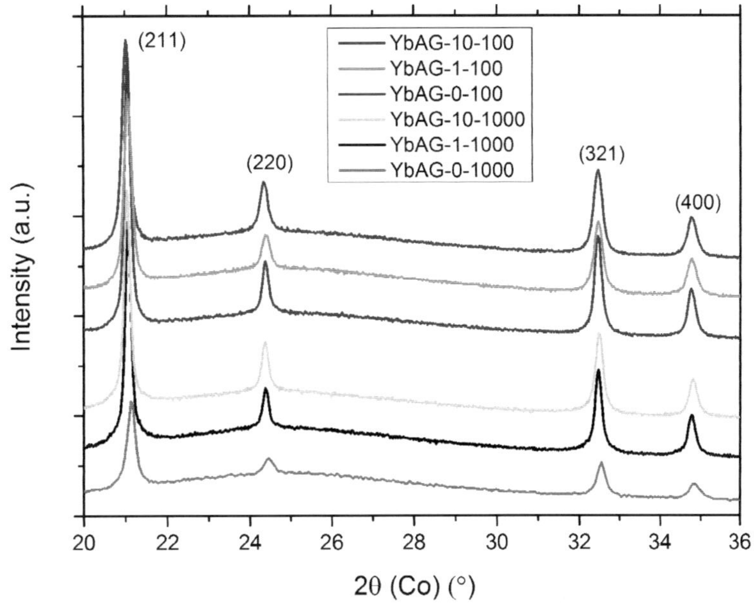

Figure 1. XRD patterns of Er^{3+}:YbAG samples in a range of $2\Theta = (20–36)°$.

Collected XRD data were used to calculate the crystallite size and internal stress of the prepared films using the Williamson–Hall equation (selected plots are shown in Figure 2); the results are summarized in Table 2. The crystallite size of all samples lies between approx. 50 nm and 120 nm. The TEOS-free samples (YbAG–0–1000 and YbAG–0–100) do not differ significantly in the crystallite size. However, the combined effect of TEOS addition and a lower pressure during the crystallization can be clearly seen – TEOS-containing samples showed lower crystallite size if crystallized at lower pressure. The internal stress values do not vary much; no conclusive relation between the internal stress on one side and the TEOS addition and/or lower pressure on the other side can be inferred.

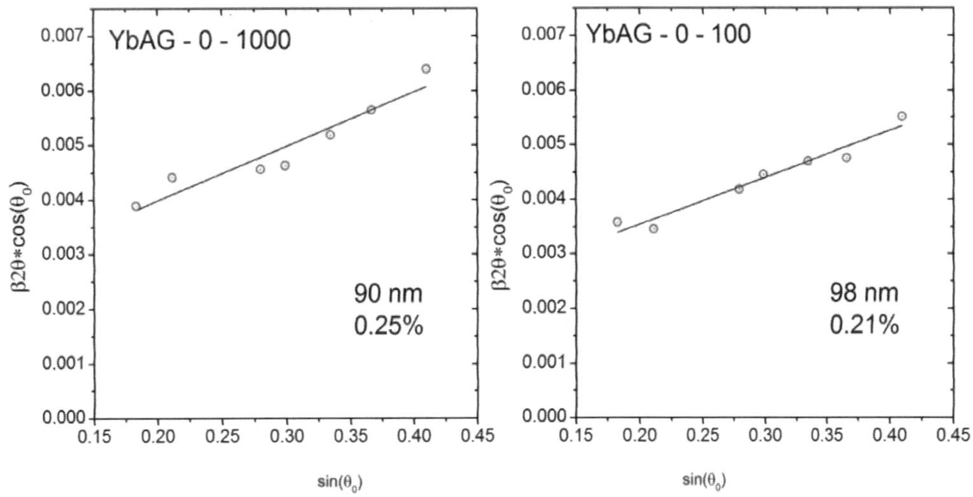

Figure 2. Williamson–Hall plots of selected samples.

Table 2. Values of crystallite size and internal stress, obtained from the Williamson–Hall plots.

Sample	Crystallite size (nm)	Internal stress (%)
YbAG–10–100	49	0.18
YbAG–1–100	62	0.27
YbAG–0–100	98	0.21
YbAG–10–1000	71	0.23
YbAG–1–1000	114	0.27
YbAG–0–1000	90	0.25

The results of the Williamson–Hall analysis were further confirmed by AFM method. Figure 3 shows the AFM images of selected samples together with the values of their roughness R_a and R_q calculated from the AFM images for all samples. TEOS-free films showed only a small change in their respective roughness (YbAG–0–1000: R_a = 6.30 nm; YbAG–0–100: R_a = 5.81 nm) when the crystallization pressure had been decreased from 1000 bar to 100 bar (images not presented). When crystallized at ambient pressure, samples with the addition of TEOS showed increase in roughness. In contrast, the roughness of the samples with TEOS (both 1 and 10 %) decreased when a lower pressure was used during crystallization of the films. This decrease is more pronounced when higher amount of TEOS was added (to approx. 1/3 of the value at ambient pressure). These results are consistent with those obtained from the Williamson–Hall plot (Table 2).

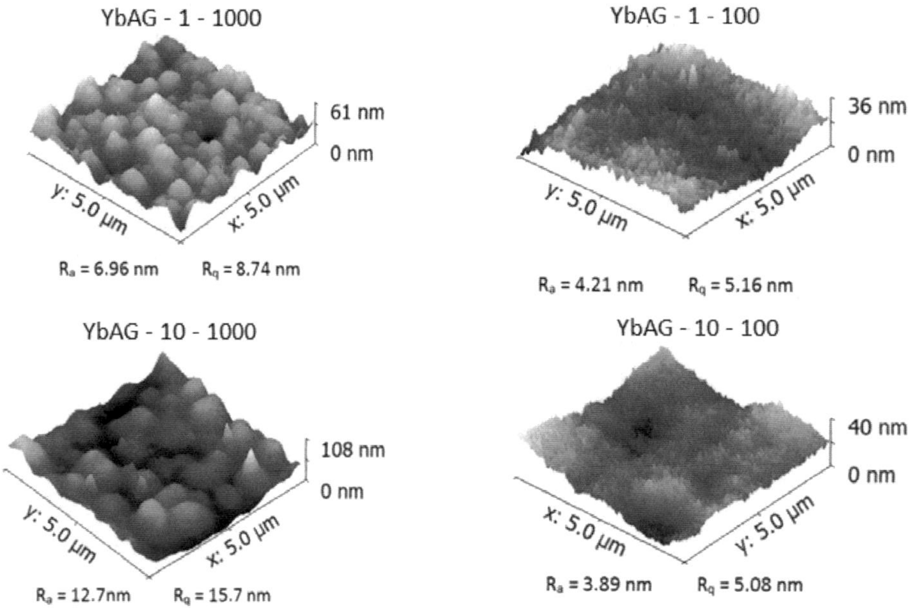

Figure 3. AFM images of selected Er:YbAG films with calculated values of roughness.

The effect of TEOS addition and a lower pressure on the surface microstructure was further observed by SEM. Figure 4 shows surface images of the films. The surface of the TEOS-free films (YbAG–0–1000, YbAG–0–100) showed no visible cracks and visually these two films have similar structure. In contrast, samples with 1 % of TEOS (YbAG–1–1000, YbAG–1–100) are visibly different. The film prepared at atmospheric pressure contained cracks, while the film with identic composition, only prepared at lower pressure (100 mbar) exhibited a smooth surface with no visible defects. Samples with 10 % TEOS added (YbAG–10–1000,YbAG–10–100) showed visible cracks over the entire surface. When putting together the results from AFM and SEM analyses, we can conclude that the combined effects of a lower pressure at crystallization and addition of small amount of TEOS led to a film with smaller crystallite size and a crackless surface. However, when the amount of added TEOS exceeded a certain level, the surface microstructure started to exhibit cracks, in spite of maintaining small crystallite size. This fact may be due to quick evaporation of volatile products from the thermal decomposition of TEOS that caused the mechanical damage of the surface.

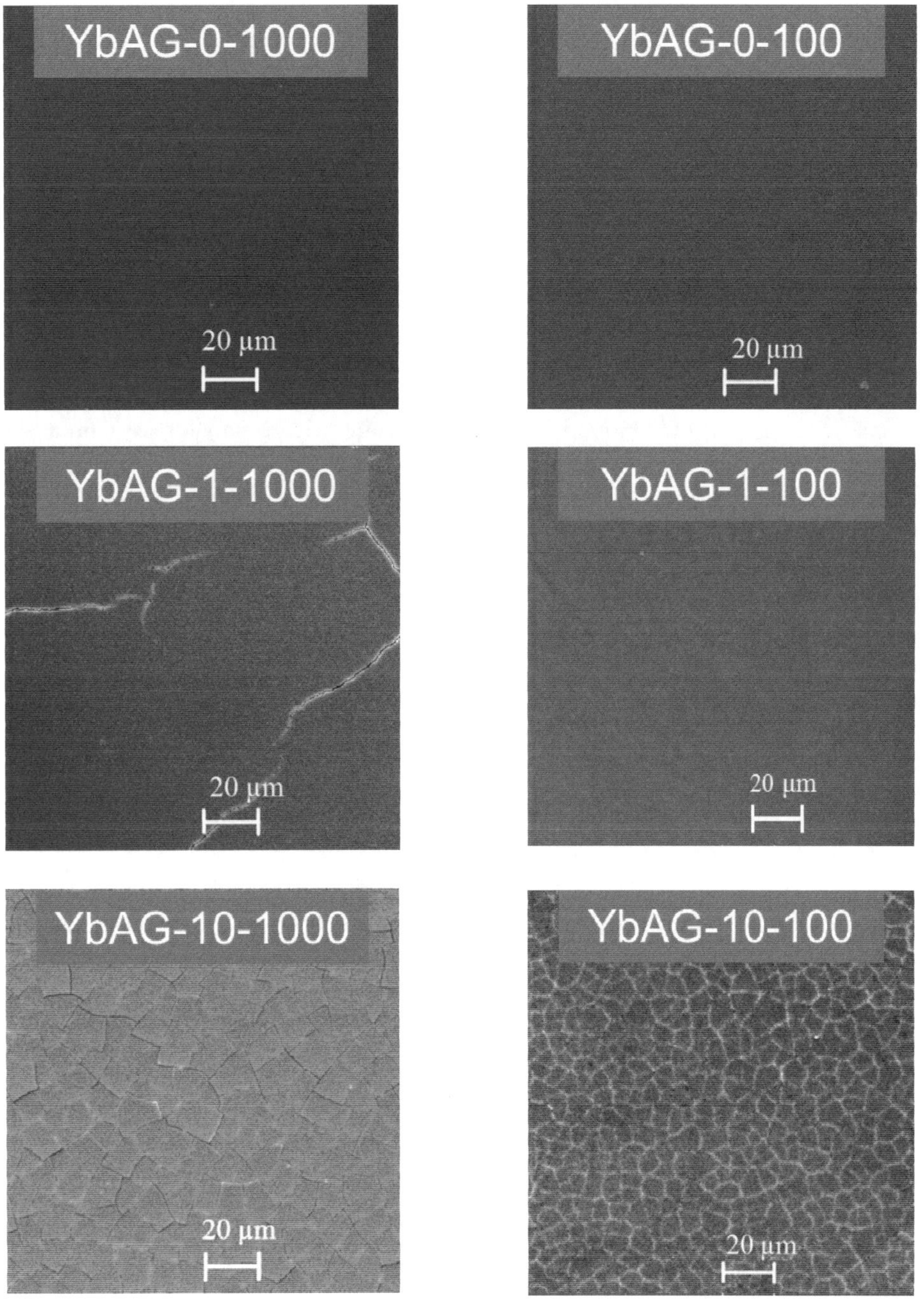

Figure 4. Scanning electron micrographs the Er^{3+}:YbAG films.

The Er doping level in samples presented here was established already in our earlier work where we found that 2 % of the ytterbium sites being substituted by erbium resulted in major intensity of photoluminescence at ~1530 nm. To see how the addition of TEOS and a lower pressure at crystal-

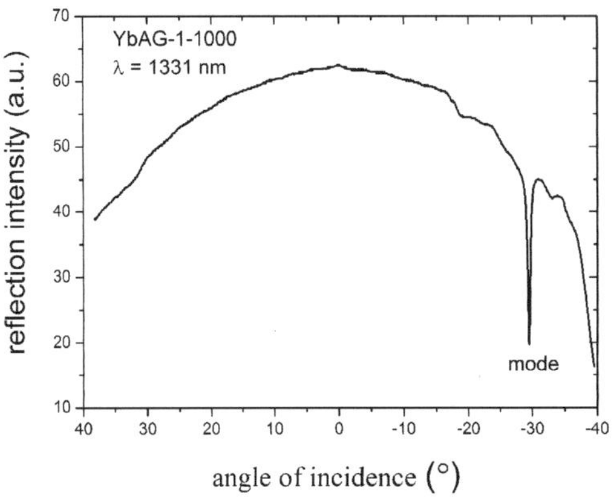

Figure 5. M-line spectroscopy record of YbAG–1–1000.

lization affected the waveguiding properties of the films, we subjected them to the m-line mode spectroscopy. A representative of this measurement is shown in Figure 5, all results are summarized in Table 3. Except for the TEOS-free films, all the other films propagated 1 mode at higher wavelengths (1331 and 1552 nm). A TEOS-modified microstructure with a smaller crystallite size apparently enhanced the waveguiding ability of the films.

Table 3. Number of guided modes in dependence of the wavelength.

Sample	Wavelength (nm)			
	633	964	1311	1552
YbAG–0–1000	0	0	0	0
YbAG–0–100	0	0	0	0
YbAG–1–1000	0	0	1	1
YbAG–1–100	0	0	1	1
YbAG–10–1000	0	0	1	1
YbAG–10–100	0	0	1	1

4. Conclusion

Three solutions for the deposition of Er^{3+}:YbAG were prepared by a wet route method based on the water soluble polymer polyvinylpyrrolidone (PVP); the solutions contained different amount of TEOS (0, 1, 10 wt%). Er:YbAG thin films were spin-coated on fused silica substrates and the effect of TEOS and lowered pressure at crystallization (100 mbar) on the microstructure was studied.

All samples were polycrystalline single phase ytterbium aluminium garnet. Using the Williamson–Hall equation, the crystallite size was calculated to be approx. 20–120 nm in all samples.

The combination of TEOS addition and decreased pressure during crystallization of the films had a beneficial effect on the microstructure (smaller crystallite size and roughness, defects-free surface). A higher amount of TEOS, however, caused cracks in the films surface regardless the crystallization pressure.

Unlike the TEOS-free samples, all films with TEOS were one-mode waveguides. Together with our previous study, this paper shows a way to improve the microstructure of waveguiding films by the use of a sintering agent.

Acknowledgment

This work was financially supported from specific university research (MSMT) No. 20-SVV/2017 and by the GA CR, project No. P108/12/G108.

References

[1] Potdevin A, Chadeyron G, Briois V and Mahiou R 2011 *Mater Chem Phys* **130** 500–6

[2] Xu C F, Yang Q B, Ren G Z and Liu Y X 2010 *J. Alloy. Compd.* **503** 82–5

[3] Li Z P, Dong B, He Y Y, Cao B S and Feng Z Q 2012 *J. Lumin.* **132** 1646–48

[4] Rubešová K, Hlásek T, Jakeš V, Matějka P, Oswald J and Holzhauser 2014 *J. Sol-Gel Sci. Technol.* **70** 142–8

[5] Cantelar E and Cussó F 2003 *J. Lumin.* **102** 525–31

[6] Shim J B, Yoshimoto N, Yoshizawa M and Yoon D H 2001 *Cryst. Res. Technol.* **36** 1209–14

[7] Salhi R, Jimenez C, Deschanvres J L, Maalej R and Fourati M 2015 *Chem. Vap. Deposition* **21** 26–32

[8] Kaczkan M, Borowska M, Malinowski M, Lukasiewicz T and Kolodziejak K 2009 *Phys Status Solidi* B **246** 1677–85

[9] Diaz–Torres L A, De la Rosa E, Salas P, Angeles–Chavez C, Arenas L B and Nieto J 2005 *Opt. Mater.* **27** 1217–20

[10] Hlásek T, Rubešová K, Jakeš V, Jankovský O and Oswald J 2014 *J. Eur. Ceram. Soc.* **34** 3779–82

[11] Zhang X, Liu H, He W, Wang J, Li X and Boughton R I 2005 *J. Cryst. Growth* **275** e1913–17

[12] Fujioka K, Saiki T, Motokoshi S, Fujimoto Y, Fujita H and Nakatsuka M 2009 *Ceram. Int.* **35** 2393–99

[13] Ramanujam P, Vaidhyanathan B, Binner J, Anshuman A and Spacie C 2013 *Ceram. Int.* **40** 4179–86

[14] Chen J, Huang X, Wang L and Zhang Q 2011 *J Rare Earth* **29** 44–7

[15] Hostaša J, Matejíček J, Nait–Ali B, Smith D S, Pabst W and Esposito L 2014 *J. Am. Ceram. Soc.* **97** 2602–06

[16] Chen X T, Lu T C, Wei N A, Lu Z W, Chen L J, Zhang Q H, Cheng G and Qi J Q 2015 *J. Alloy. Compd.* **653** 552–60

[17] Kong L B, Huang Y Z, Que W X, Zhang T S, Li S, Zhang J, Dong Z L and Tang D Y 2015 Sintering and densification of transparent ceramics *Transparent Ceramics (Topics in Mining, Mettallurgy and Materials Engineering)* ed. C P Bergmann (Cham: Springer International Publishing) chapter 7 pp 467–517

[18] Rubešová K, Mikolášová D, Hlásek T, Jakeš V, Nekvindová P, Bouša D and Oswald J 2016 *J. Lumin.* **176** 260–65

[19] Hlásek T, Polák V, Rubešová K, Jakeš V, Nekvindová P, Jankovský O, Mikolášová D and Oswald J. 2016 *J. Sol-Gel Sci. Technol.* **80** 531–37

[20] Williamson G K and Hall W H 1953 *Acta Metall.* **1** 22–31

IOP Conf. Series: Materials Science and Engineering **266** (2017) 012005 doi:10.1088/1757-899X/266/1/012005

Unusual Attenuation Recovery Process After Fiber Optic Cable Irradiation

Z Konečná[1], V Plaček[2] and P Havránek[2]

[1] Czech Technical University in Prague, Department of Electrotechnology, Technická 2, Praha, Czech Republic
[2] ÚJV Řež, a. s., Hlavní 130, Řež, Husinec, Czech Republic

zuzana.konecna@fel.cvut.cz

Abstract. At present, the number of optical cables in nuclear power plants has been increasing. Fiber optic cables are commonly used at nuclear power plants in instrumentation and control systems but they are usually used in environments without radiation. Nevertheless, currently, the number of applications in NPP containment with radiation is increasing.
One of the most prevalent effects of radiation exposure is an increase of signal attenuation (signal loss). This is the result of fiber darkening due to radiation exposure and it is the main limitation factor in application of fiber optics in radiation environment. However, after the irradiation, the fiber optics go through a "recovery process" during which the optical properties improve again; i.e. attenuation decreases. However, we have found cable, where the expected healing process after few days changed its trend and the attenuation increased again to a value well above the attenuation just after the irradiation. This paper describes experiments that were carried out to explain this unusual recovery behaviour.

1. Introduction

Application of safety related fiber optic cables in harsh environments of nuclear power plants (NPPs) requires their qualification. The international standard for qualification IEEE 1682: Standard for Qualifying Fiber Optic Cables, Connections, and Optical Fiber Splices for Use in Safety Systems in Nuclear Power Generating Stations [1] has existed since 2011. This standard supplements more general qualification standard IEC/IEEE 60780-323 [2].

In nuclear research institute ÚJV Řež laboratory, we tested and qualified fiber optic cables for NPP applications via simulation of degradation factors of optical cables by accelerated ageing.

Main degradation factors are:

- Temperature
- Mechanical stress
- Humidity
- Radiation

All these factors have a negative influence on polymer materials used in jackets and tubes, which are studied and known thank to metallic cables. Less information exists about ageing of fiber optic glass and the importance of individual stressors above mentioned. Long term effect of temperature and humidity may lead to embrittlement of glass, cause microcracks and diffusion of OH ions into glass structure.

Content from this work may be used under the terms of the Creative Commons Attribution 3.0 licence. Any further distribution of this work must maintain attribution to the author(s) and the title of the work, journal citation and DOI.
Published under licence by IOP Publishing Ltd

IOP Conf. Series: Materials Science and Engineering **266** (2017) 012005 doi:10.1088/1757-899X/266/1/012005

The most important stressor in NPP is ionizing radiation, which causes fiber darkening, i.e. dramatically attenuation increase and loss of functionality. Need to be mentioned, that the total dose causing glass to be "blind" is much lower than the dose, which would case other fiber optic cable polymer materials degradation.

The influence of ionizing radiation initiates significant changes in glass matrix. Photons of gamma-radiation infiltrate deep into the glass structure and in the simplest case cause oxidation-reduction chemical processes [4], but most serious effects of high-energy gamma radiation include a number of changes in chemical, mechanical and optical, properties of glass. Electrons, released by ionization, can be trapped in grid structure creating defects that change temporarily or permanently the glass structure.

Deflections resulting from the displacement of atoms (atoms or ions are moved from their equilibrium into interstitial positions) are accompanied by energetic level changes of their electrons. If the ion leaves its equilibrium position, the vacancy acts as positive charged. The electrons released from the glass due to the radiation, are captured here and create a color centers with typical absorption in the visible area of the spectrum [5]. The direct consequence is the change in the absorption characteristics leading to the coloration of the fiber (glass goes through the darkening process) and hence attenuation increases. This is so significant that it can lead to the "blinding" of optical fiber with attenuations in tens of dB per km of its length. Such a value can affect proper function of fiber optic cables in NPP safety systems.

Qualification according to the standard IEEE 1682 [1] requires following steps: functionality testing, accelerated thermal ageing, accelerated radiation ageing using irradiation of cables with gamma rays, coiling the cables around mandrel to simulate mechanical stress encountering from the operation and checking the functionality during the recovery process after irradiation, finally accident condition simulation with functionality testing.

A new installed cable shall meet all the requirements described in its technical specifications and/or requirements established in the respective NPP design. Nevertheless, during the service ageing the cable may change some of its properties but it shall not be damaged to the extent which might compromise its proper function. Following radiation exposure the light transfer function of optical cable fibers will be ensured if the change of attenuation is smaller than 10dB/100m. For telecommunication applications, this value would be not acceptable. But for NPP safety systems, where the total length is typically bellow 100 m this value can be used as an end-point acceptance criterion.

2. Influence of Gamma-irradiation on Fiber Optic Cable

As mentioned above, increasing attenuation and especially the "recovery process", are the most important factors that limits fiber operability. The extent of recovery can be affected among the others by time and temperature after the irradiation [6]. But it will never come back to the initial value. This process and description how to take it into account during qualification for nuclear power plant application are described in the IEEE 1682 standard [1].

2.1. Recovery Process - Relaxation

We started our activities in NPP fiber optic qualification with simplex and multicore fiber optic cables. The used fibers were multi-mode (MM) fibers with graded index, consisting of a fiber optic core covered with a buffer layer. The used fiber was: MM 62.5/125 radiation hardened fiber. The buffer was surrounded by aramid yarn and a final jacket. Cables were irradiated with gamma rays from ^{60}Co source to the total dose 1.7 kGy at the dose rate 40 Gy/h. After the end of the irradiation, the cable was placed in a room with controlled temperature for the period up to 30 days to check attenuation recovery. Results are plotted in Figure 1 and Figure 2. The first figure shows the attenuation increase during the irradiation, while Figure 2 represents a 10 days recovery process at room temperature. It is a typical and expected behavior [3]; attenuation increases during irradiation with partial recovery after the irradiation.

Quite different was the situation for jelly filled loose tube single-mode (SM) cable. Cable was during the qualification irradiated with the dose 15 kGy at the dose rate 0.3 kGy/h. The attenuation increased to quite high value. When the irradiation was stopped, the attenuation as expected decreased (recovered).

Nevertheless, the recovery took only 3 days. After this short period, the trend changed and attenuation increased to a value well above the attenuation just after the irradiation; well above the acceptance criterion for NPP application. For comparison see expected (model) vs. real conditions in Figure 3. We have called this unusual behavior as "inverse recovery process".

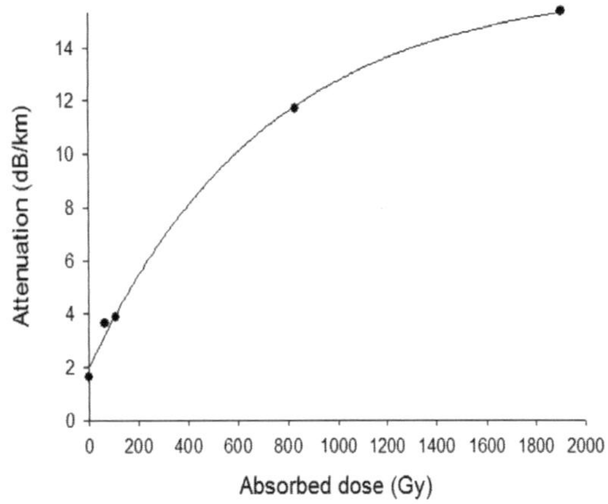

Figure 1. Attenuation increases during irradiation measured on MM simplex cable. Measuring was performed on wavelength 1300 nm in radiation interruptions.

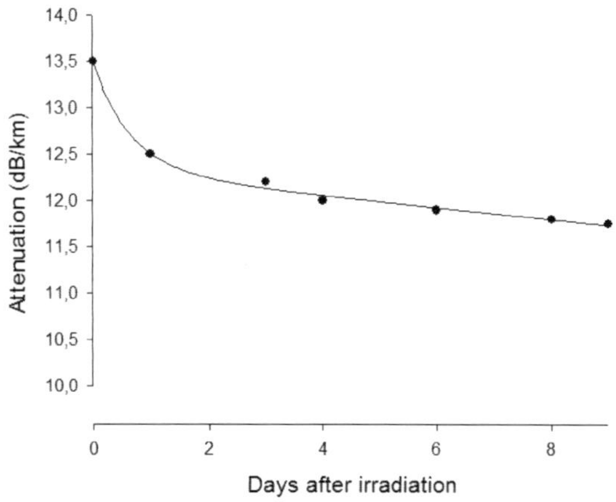

Figure 2. Recovery process initiation measured on MM simplex cable mentioned in Figure 1. Measured at 1300 nm. Measurement started 2 hours after irradiation.

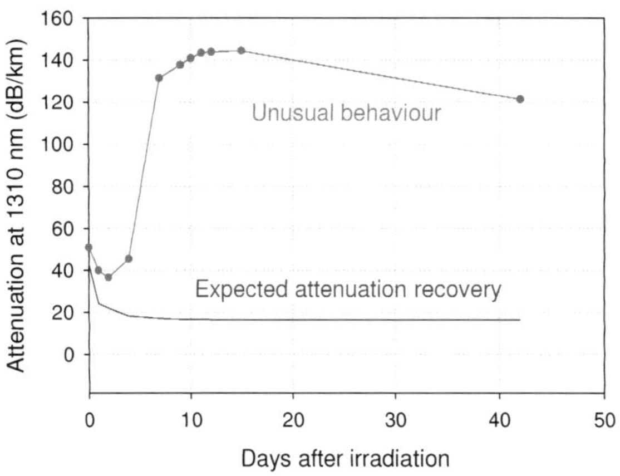

Figure 3. Attenuation recovery after fiber optic cable irradiation measured at 1310 nm. Expected is attenuation decrease, i.e. improvement of optical properties. Some types of fiber optic cables exhibits an unusual attenuation increase after few days of recovery time.

2.2. Reason of Observed Phenomenon
There are at least two main processes, attenuation recovery process, which is typical for irradiated Si glass and some process, quite slow at the beginning but that tends to dominate after few days. Possible explanations and reasons can be following:

1. Problem is in connectors and not in the cable itself.
2. Ge dopant in the SM fiber.
3. Thermal ageing that was carried out before irradiation
4. Mechanical stress. After irradiation, the cables must be, in accordance with the standard [1], straightened and coiled around a cylinder with a diameter similar to maximal allowed bending radius.
5. Dose rate of irradiation. During the operation, the cables are irradiated at very low dose rates, while irradiation in laboratory uses much higher ones. Is there any dose rate effect?
6. Total dose. MM cable was irradiated with ten times lower dose.
7. Different cable construction, used materials, fibers (MM vs. SM), tubes, jelly etc.
8. Some stress encountering during the manufacturing process.
9. Other.

3. Methodology of Experiment
3.1. Attenuation Measurement
As measuring method for transparence fiber glass detection, measuring of attenuation was chosen. MM cables were measured on wavelengths 850 nm and 1300 nm and SM cables on 1310 nm and 1550 nm (in few cases on 1625 nm).

During experiments, we used two attenuation – measuring methods:

- LSPM – Light Source and Power Meter (EN/IEC 60793-1-40 – method B Insertion Loss)
- OTDR – Optical Time Domain Reflectometer (EN/IEC 60793-1-40 – method C Backscattering)

It needs to be noticed, that all these experiments require quite long cable samples. The measuring devices need to be connected to the tested samples during 30 days of the recovery process and cannot be easily used for another experiment. Therefore, it was performed only limited amount of experiments and some of our conclusions were not proved on more samples. The most important was for us to find what causes the "inverse recovery process".

Attenuation measurement was performed discretely in the time for trend monitoring or continuously.

3.2. Irradiation
To measure attenuation with a reasonable uncertainty, cables need to be at least 100 m long. To increase the optical length of the cable, individual fibers within the cable can be connected to series by fusion. Nevertheless, it must be ensured, that these fusions and connections do not influence the results.

Sample is attached to a perforated stainless steel cylinder with the diameter of 0.4-0.8 m, while the cobalt source is in the middle of the cylinder. Cable ends with connections and protective boxes are placed to a position with much lower dose rate or they are shielded. After irradiation, cables are straightened and bent around a mandrel to minimal allowed bending radius for installations, which was for the tested cables 20 × cable diameter (20D), such a mechanical stress is required by the standard [1]. Samples are kept at room temperature. Attenuation is measured during the recovery period.

4. Results and Discussion
We have performed a lot of experiments with the aim to explain this inverse recovery process. Firstly, we checked individual steps of the qualification process. In the first step, we focused on explanation of possible cable connection effect, fiber itself, thermal ageing which had to be applied before irradiation, mechanical stress and irradiation process. Table 1 summarizes experiments carried out to explain inverse recovery process.

Table 1. Experiments overview.

Experiment	Sample type	Note
Connector and fusions irradiation	Whole cables	
Fiber dopants	Fibers	Two types of used dopants
Thermal ageing prior to irradiation	Whole cables	
Mechanical Stress after Irradiation	Fibers and cables	Influence of recoiling after irradiation exposure
Dose Rate Effect	Fibers and cables	

4.1. Connector and fusions irradiation

To check the possible influence of connectors and fusions, same samples were irradiated in such a way, that the cables ends were shielded and/or pulled out of the irradiation facility. There was no difference between the ends with connectors that were pulled out (i.e. not irradiated) or were slightly irradiated which was the case of previous testing.

4.2. Fiber dopants

To check if the germanium (Ge) plays a role in attenuation increase after irradiation, fibers with two different dopants were tested: fiber with Ge dopant, Single mode Fiber SM (G652D); and Single mode Fiber SM, radiation hardened fiber, unknown dopants (but no Ge).

Both fibers were irradiated with the dose 15.7 kGy at the dose rate 1.7 kGy/h. Samples were attached to a cylinder with the diameter 40 cm. Attenuation after the irradiation is in Figure 4.

These experiments gave very important results:

- There is no attenuation increase during recovery process if only fiber alone is irradiated. Compare Figure 3 and 4: in both experiments the same Ge fiber was irradiated. Figure 3 – whole cable; Figure 4 – only fiber alone.
- No effect of the used dopant (Figure 4).

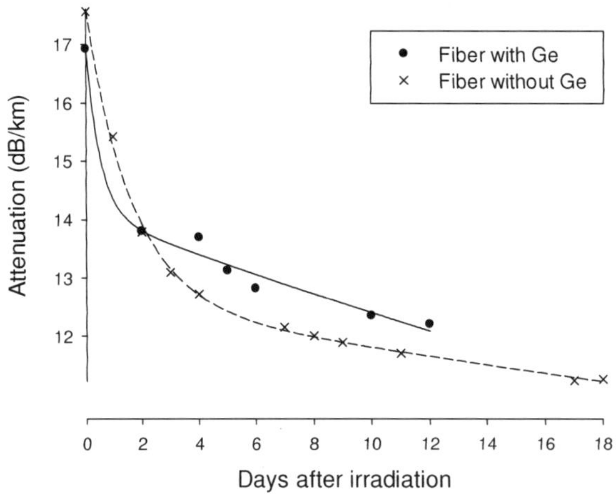

Figure 4. Monitored influence of used dopants in fiber structure. Fiber with Ge is optic fiber with germanium dopant. Fiber without Ge is optical fiber with another unknown dopant.

4.3. Thermal ageing

First step of qualification is simulation of long term thermal ageing. In accordance with the standard [1], thermal ageing is simulated using Arrhenius approach, which assumes that a short–term simulation of thermal ageing at higher temperature causes the same degradation as long–term ageing at lower temperatures. Cable was aged at 100 °C for the period of 110 days before it was irradiated. To check the influence of possible influence of thermal ageing, cable was irradiated also without any thermal treatment. Both cables, thermally aged as well as without thermal treatment exhibited the same inverse recovery process.

4.4. Mechanical Stress after Irradiation

Mechanical stress has the origin in sample straightening and bending immediately after the irradiation. Such a procedure is required by the standard and simulates possible mechanical stresses encountering in the NPP operation and during some accidents.

Therefore, cables coiled around a mandrel after irradiation as well as samples without mechanical stress were measured during the recovery process. There was no difference; i.e. the mechanical stress is not the reason for inverse recovery behavior.

Additionally, two samples of only SM fiber were irradiated together on a drum of 40 cm diameter. After the irradiation one fiber was kept on the drum without any manipulation, while the second one was very carefully straightened and recoiled on a smaller drum with the diameter 10 cm. Attenuation on both samples was measured during the recovery time. Results are displayed in Figure 5. Mechanically treated has little bit higher attenuation. Nevertheless, the effect is small, there is almost negligible difference between both samples.

Summary of the mechanical stress:

- Mechanical stress caused by the cable straightening and subsequent bending on small radius does not influence the process of the recovery and its inverse behavior after the irradiation.
- We eliminated mechanical stress as a cause of induced microbendings. We checked the OTDR traces and there is not such a big difference between attenuation measured at 1310 nm and 1550 nm, even 1625 nm. Microbending losses shall decreases at longer wavelengths [6], but we did not observed this trend.

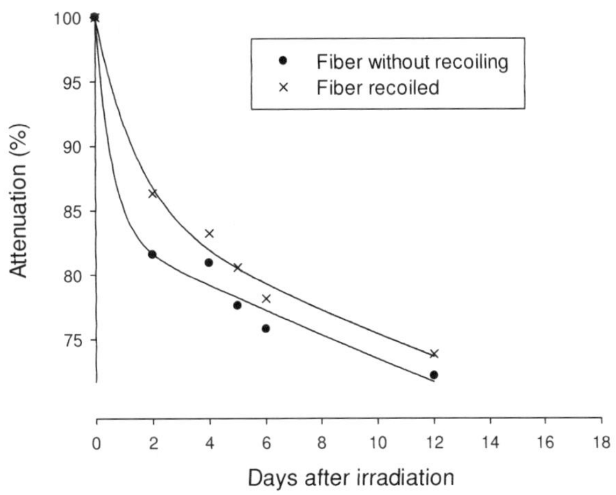

Figure 5. On y-axis is relative change of attenuation (100 % is value immediately after irradiation). Fiber recoiling was left on the drum with diameter of 40 cm without any manipulation while the second sample was recoiled on a smaller drum with the diameter of 10 cm.

4.5. Dose Rate Effect

A loose tube SM jelly filled loose tube single-mode (SM) cable was irradiated at the dose rate 1.45 kGy/h and compared with the same cable irradiated at 0.0173 kGy/h, i.e. approx. 100 times lower dose rate. Total dose was around 15 kGy. Difference in the recovery process is quite visible. Even the attenuation just after the irradiation at both dose rates was similar, at low dose rate irradiation is the inverse recovery process much less important, see Figure 6.

To check the attenuation increase after irradiation at different dose rates on fibers used in the cable, only fibers alone were irradiated at the same conditions as mentioned above.

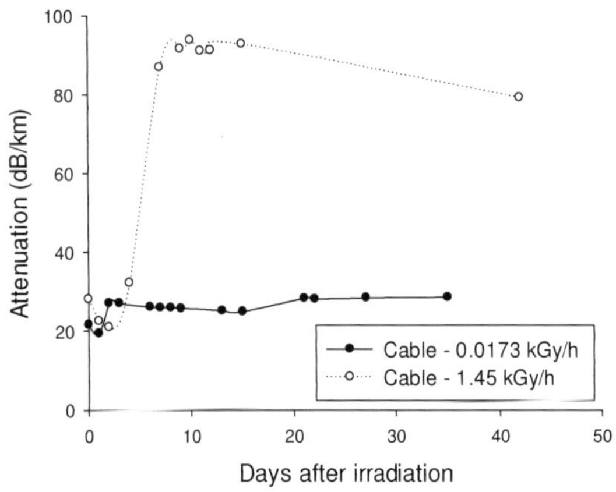

Figure 6. Influence of dose rate irradiation of whole cable on the attenuation and recovery process. Total dose was 15 kGy.

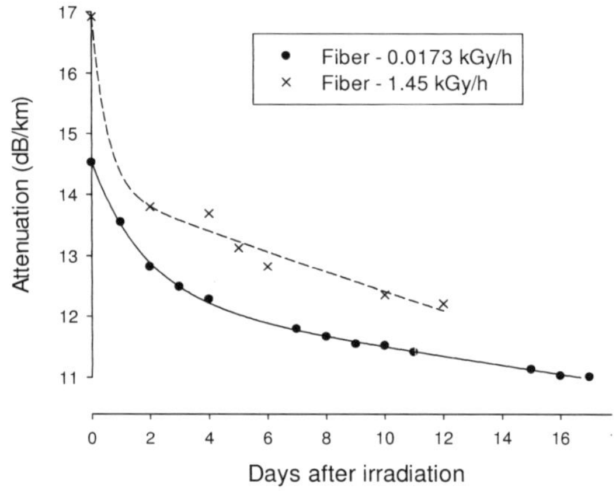

Figure 7. Influence of dose rate on recovery of fiber samples, that were irradiated alone.

The attenuation increase due to the irradiation, was not influenced by the applied dose rate significantly; i.e. the attenuation immediately after the irradiation was almost the same when irradiated fast at 1.45 kGy/h or slowly at .0173 kGy/h (see Figures 6 and 7). However, important fact is, that the attenuation values after irradiation were higher for cable samples than for fibers. At higher dose rate, it was approx. two times, at small dose rate only one and half times.

Summary of the dose rate experiments:

- The applied dose rate is not important for final attenuation increase during and immediately after irradiation.
- Inverse recovery effect was much higher when irradiated at higher dose rate.
- Attenuation increase during and after irradiation was approx. two times higher when irradiated whole cable comparing to irradiation only fiber alone.

5. Conclusion

While we tested fiber optic cable for the use in NPPs, we discovered unusual phenomenon "inverse recovery process". Expected behavior of fiber optic cable after irradiation expose is improving of optical properties; i.e. glass fiber become darker while irradiated but during the recovery process without radiation, the optical properties improve. However, for some cables, improving of their properties after irradiation stopped after short period and attenuation started to increase to a value well above the values measured just after the irradiation. We performed a lot of experiments to explain this unusual phenomena. At first, we tested the possible influence of individual steps of the fiber optic cable qualification process. Results of our experiments could, as the reason for inverse recovery process, exclude the influence of thermal ageing before irradiation, mechanical stress encountering from the bending of irradiated cable around mandrel as well as the applied dose rate. Moreover, attached connectors and fusions on qualified cable were excluded as well. Another set of experiments checked possible effect of dopants in the glass fiber. They showed, that there is no effect of the fiber dopant.

Interesting results appeared when we compared the recovery process of whole irradiated cable with only fiber. There was no attenuation increase (inverse recovery process) when only fibers alone were irradiated. The root cause of the inverse recovery process phenomenon is not known yet. However, we suppose that this effect is limited to the cable construction. Thus we are planning experiments with other materials of tubes, analyzing role of coating, jelly, influence of fiber amount in tube etc.

Acknowledgments
Test samples were kindly provided by GOC Co.,Ltd (Fiber optic cable manufacturing company, Gwangju, Korea). This work was supported by the Grant Agency of the Czech Technical University in Prague, grant SGS 16/224/OHK3/3T/13. Special thanks to colleagues from VŠB (Technical University of Ostrava, Faculty of Electrical Engineering and Computer Science, Department of Telecommunications), for the initial cooperation and measurement.

References
[1] IEEE1682-2011: Standard for Qualifying Fiber Optic Cables, Connections, and Optical Fiber Splices for Use in Safety Systems in Nuclear Power Generating Stations
[2] IEC/IEEE 60780-323:2016: Nuclear facilities – Electrical Equipment important to safety – Qualification
[3] Vašínek V, Hájek L, Bednárek L, Vanderka A, Hlavinka T and Marcinka O 2015 Influence of gamma radiation on optical fibers and devices *SPIE 9659*
[4] Fanderlík I 1996 Glass Properties "Vlastnosti skel" 1st ed, *Informatorium*
[5] Sandhu A K, Singh S and Pandey O P 2008 Gamma ray induced modifications of quaternary silicate glasses *J. Phys. D: Appl. Phys.* **41** 165402
[6] Senior J M and Jamro M Y 2009 Optical Fiber Communications: Principles and Practice 3rd ed, *Pearson Education*
[7] Griscom D L, Gingerich M E and Friebele E J 1994 Model for the dose, dose-rate and temperature dependence of radiation-induced loss in optical fibers *IEEE Trans. Nucl. Sci.* **41** 523–527
[8] Girard S, Kuhnhenn J, Gusarov A, Brichard B, Van Uffelen M, Ouerdane Y, Boukenter A and Marcandella C 2015 Radiation Effects on Silica-Based Optical Fibers : Recent Advances and Future Challenges *IEEE Trans. Nucl. Sci.*, **60** 2015–36

IOP Conf. Series: Materials Science and Engineering **266** (2017) 012006 doi:10.1088/1757-899X/266/1/012006

Size distribution and growth rate of crystal nuclei near critical undercooling in small volumes

Z Kožíšek and P Demo

Institute of Physics of the Czech Academy of Sciences, Cukrovarnická 10, 162 00 Praha 6, Czech Republic

kozisek@fzu.cz

Abstract. Kinetic equations are numerically solved within standard nucleation model to determine the size distribution of nuclei in small volumes near critical undercooling. Critical undercooling, when first nuclei are detected within the system, depends on the droplet volume. The size distribution of nuclei reaches the stationary value after some time delay and decreases with nucleus size. Only a certain maximum size of nuclei is reached in small volumes near critical undercooling. As a model system, we selected recently studied nucleation in Ni droplet [J. Bokeloh et al., Phys. Rev. Let. 107 (2011) 145701] due to available experimental and simulation data. However, using these data for sample masses from 23 μg up to 63 mg (corresponding to experiments) leads to the size distribution of nuclei, when no critical nuclei in Ni droplet are formed (the number of critical nuclei < 1). If one takes into account the size dependence of the interfacial energy, the size distribution of nuclei increases to reasonable values. In lower volumes ($V \leq 10^{-9}$ m^3) nucleus size reaches some maximum extreme size, which quickly increases with undercooling. Supercritical clusters continue their growth only if the number of critical nuclei is sufficiently high.

1. Introduction

Nucleation (homogeneous or heterogeneous) is the initial process leading to the formation of a new phase when a certain nucleation energy barrier is overcome. In the case of homogeneous nucleation nuclei of a new phase are formed at any monomer (atom or molecule) within parent phase while at heterogeneous nucleation the nuclei are created on foreign surfaces, impurities, surface defects, etc. A special case of heterogeneous nucleation is formation of nuclei on active centers [1-3], when its probability on these centers is higher. Nucleation plays a decisive role in determining the crystal structure and the size distribution of nuclei. A better understanding of nucleation opens the possibilities to new approaches to control of crystallization process.

Critical undercooling (corresponding to situation when a detectable amount of thermodynamically stable phase is formed), induction time, and metastable zone width provide important information about nucleation process. From the viewpoint of nucleation theory the computation of the critical undercooling (resp. supersaturation for the formation of droplets form vapor or crystals from solution) is based on assumption that the stationary nucleation rate reaches a certain concrete value [4]. Alternatively, the critical supersaturation can be derived from the condition that one supernucleus of preset number of molecules (or atoms) is formed in a parent phase [5, 6].

Repeated measurement of the critical undercooling showed that nucleation is a stochastic process [7, 8]. If one supposes Poisson distribution of nuclei [9-11], the nucleation rate and the work of

Content from this work may be used under the terms of the Creative Commons Attribution 3.0 licence. Any further distribution of this work must maintain attribution to the author(s) and the title of the work, journal citation and DOI.
Published under licence by IOP Publishing Ltd

formation of nuclei can be derived from the survivorship function defined as the fraction of experiments in which crystallization did not yet occur at a given value of undercooling ΔT. Recently, Bokeloh et al. [7, 12] analyzed the differential scanning calorimetry (DSC) experiments to determine nucleation rate and the work of formation of nuclei in Ni droplets together with the Monte Carlo (MC) simulations. In DSC experiments the released latent heat is detected. However, one has no information about the size distribution of nuclei and the maximum size of nuclei formed within nucleating system.

The aim of this paper is to extend aforementioned research to determine the size distribution of nuclei by numerical solution of kinetic equations in small Ni droplets (for which all input data are known) and to show how small changes in critical undercooling influence the maximum size of formed nuclei.

2. Model

Homogeneous nucleation in Ni droplets is considered within context of the standard nucleation theory. The number of nuclei, $F_i\,(t)$, formed by i atoms at time t is governed by the following equation [13]:

$$\frac{dF_i}{dt} = J_{i-1} - J_i \tag{1}$$

where

$$J_i(t) = k_i^+ F_i - k_{i+1}^- F_{i+1} \tag{2}$$

is the nucleation rate (i. e. the number of i-sized nuclei formed in unit volume at time t) and k_i^+ (k_i^-) denotes the attachment (detachment) frequency.

The attachment frequency in condensed systems was introduce by Turnbull and Fisher [4]:

$$k_i^+ = \rho_S A_i \left(\frac{k_B T}{h}\right) \exp(-\frac{E}{k_B T}) \exp\left(-\frac{q\,\Delta g_i}{k_B T}\right), \tag{3}$$

where ρ_S denotes the number density of nucleation sites on the nucleus surface,

$$A_i = \gamma i^{2/3} \tag{4}$$

is the surface area of nucleus, E stands for the activation energy of diffusion across the phase interface, k_B and h denote the Boltzmann and Planck constants, T is temperature,

$$\Delta g_i = W_{i+1} - W_{i,} \tag{5}$$

and the work of formation of clusters using the capillarity approximation can be expressed by:

$$W_i = -i\Delta\mu + A_i \sigma. \tag{6}$$

Above,

$$q = 0.5[1 + \mathrm{sign}(\Delta g_i)] \tag{7}$$

and σ is the interfacial energy of nucleus. The critical size, i^*, corresponds to the size, when W_i reaches maximum and thus one gets (from $dW_i/dt = 0$):

$$i^* = \left(\frac{2\gamma\sigma}{3\Delta\mu}\right)^3. \tag{8}$$

Subcritical nuclei ($i<i^*$) have tendency to dissolve ($k_i^+ < k_i^-$) and supercritical nuclei have tendency to growth ($k_i^+ > k_i^-$). At the critical size both transient frequencies reach the same value (metastable equilibrium).

Molecular simulations [15, 16] showed that σ increases with i. Tolman [17] derived the following equation:

$$\sigma(i) = \sigma_\infty(1 - 2\delta/r) \tag{9}$$

for the size dependence of σ, where σ_∞ denotes the interfacial energy in the limit of flat interface. However, this dependence varies greatly with the applied value of the Tolman length δ. Merikanto et al. [15] derived the following approximation for Lenard-Jones clusters:

$$\sigma(i) \approx \sigma_\infty - \frac{A_1\sigma_\infty + 13.6\, k_B T}{A_i} \tag{10}$$

where no parameter δ is needed.

Similarly as Bokeloh et al. [7] the spherical nucleus shape was considered and thus the following relationship between nucleus radius *r* and the number of nuclei within nucleus holds:

$$\frac{4}{3}\pi r^3 \rho_C = i\, m_1, \tag{11}$$

where m_1 is the molecular mass and ρ_C is the density of crystalline phase and, consequently [see Eq. (4)],

$$\gamma = \sqrt[3]{36\pi}\left(\frac{m_1}{\rho_C}\right)^{2/3}. \tag{12}$$

The detachment frequency can be determined from the local equilibrium principle (where J_i=0):

$$k_i^+ F_i^0 = k_{i+1}^- F_{i+1}^0, \tag{13}$$

where

$$F_i^0 = N_1 exp\left(\frac{W_i}{k_B T}\right) exp\left(-\frac{W_i}{k_B T}\right) \tag{14}$$

denotes the equilibrium distribution of nuclei, N_1 is the number of atoms in undercooled melt.

The difference in the chemical potential between both phases can be approached by:

$$\Delta\mu = \frac{\Delta h_E(T_E - T)}{N_A T_E}, \tag{15}$$

where Δh_E is the heat of fusion, T_E denotes the melting temperature and N_A is the Avogadro constant.

Under assumption that the number of atoms in the parent phase does not change in time the stationary state is reached after a certain induction time, when $J_i = J^S$ = const. The stationary nucleation rate is determined by [13]:

$$J^S = \left(\sum_{i=1}^{\infty}\frac{1}{k_i^+ F_i^0}\right)^{-1}. \tag{16}$$

The growth rate of nuclei can be computed from the transient frequencies:

$$v = \frac{dr}{di}\frac{di}{dt} = \frac{dr}{dt}\left(k_i^+ - k_{i+1}^-\right) \tag{17}$$

and in the limit of flat surface ($i\to\infty$) one gets:

$$v_\infty = \frac{1}{3}\rho_S\left(\frac{k_B T}{h}\right) exp\left(-\frac{E}{k_B T}\right)\left(\frac{3m_1}{4\pi\rho_C}\right)^{1/3}\left[1 - exp\left(-\frac{\Delta\mu}{k_B T}\right)\right]. \tag{18}$$

Growth rate *v*, determined according to Eq. (17), leads to a negative values for subcritical sizes. Alternatively, one can determine the growth rate from the size of the largest nucleus i_{max} [respectively r_{max} from Eq. (11)] formed within volume V at time *t*, when

$$F_i(t)\, V = 1 \tag{19}$$

and thus $i_{max}(t)$ [respectively $r_{max}(t)$] dependence is determined.

For numerical solution of kinetic equations, the following initial conditions will be used:

$$F_i(t=0) = F_i^0 \quad \text{for } i \le i_0 \tag{20}$$

$$F_i(t=0) = 0 \quad \text{for } i > i_0, \tag{21}$$

$$F_1(t) = N_T - \sum_{i=2}^{l} F_i(t), \quad (22)$$

$$F_l(t) = 0, \quad (23)$$

where l denotes the largest size of nuclei considered in the computation and

$$N_T = N_1 + \sum_{i=2}^{i_0} i\, F_i^0 \quad (24)$$

is the total number of atoms in nucleating system. Eq. (22) takes into account decrease of atoms within parent phase. Classical nucleation theory (CNT) suppose that the number of atoms within parent phase does not change, i. e. the following condition is used:

$$F_1(t) = N_1. \quad (25)$$

3. Results and discussion

As a model system, homogeneous nucleation in Ni droplets was chosen. Following accessible experimental data and also input parameters for standard model are used: Δh_E = 17.29 kJ mol^{-1}, $T_E = 1748\ K$, ρ_C = 8357 kg m^{-3} and $\sigma = (\sigma_m T/T_E)$, where σ_m = 0.275 J m^{-2} [7]. We numerically solved Eq. (1) for maximum size of nuclei l = 250000 [see Eq. (23)] to determine the size distribution of nuclei, F_i (t), at undercooling $\Delta T = (T_E - T)$ = 300 K (corresponding to critical size i^* = 450) supposing size-independent interfacial energy – see Figure 2 in [18]. The number of critical nuclei at the stationary state $F_{450}^S \approx 4 \times 10^5\ m^{-3}$ and one gets for the volume, in which one nucleus is formed, $V_1 = 1/F_i^S = 2.5 \times 10^{-6}$ m^3. Critical supersaturation was measured for Ni sample masses from 23 μg up to 64 mg [7] which corresponds to sample volumes V = 2.9×10^{-12} m^3 – 7.15×10^{-9} m^3 (*i. e.* much less than V_1). No critical nuclei are formed according to the standard nucleation theory in the difference of experimental data. That is why we have taken into account the size dependence of the interfacial energy (Figure 1) similarly as in our previous work [18]. However, for small cluster sizes (subcritical nuclei) the σ (*i*) gives negative values for Eq. (10) [dotted line in Figure 1] and also for Eq. (9) with δ = 1.5 Å (dashed line). In further numerical computations σ (*i*) dependence was taken using Eq. (9) with δ = 0.8 Å (full line), for which σ (*i*) > 0 for any *i*.

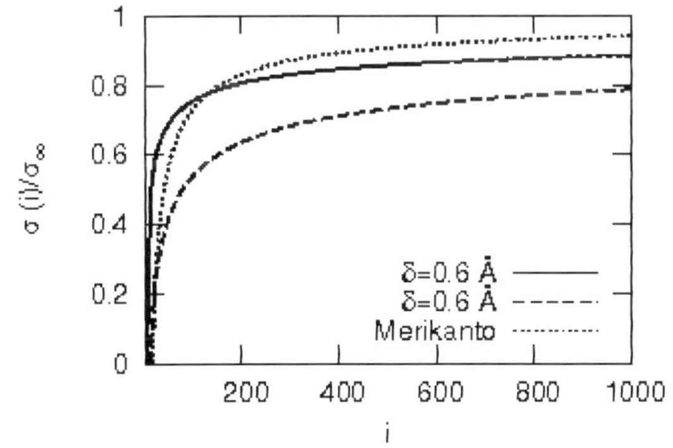

Figure 1. The interfacial energy σ (i) scaled by the flat interface limit σ_∞ as a function of cluster size *i* using Eq. (10) [dotted line] and Eq. (9) with δ = 1.5 Å (dashed line) and 0.8 Å (full line).

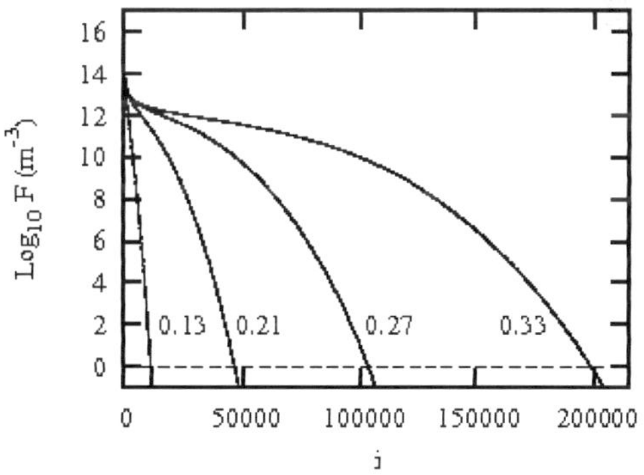

Figure 2. The size distribution of nuclei, *F*, as a function of cluster size, *i*, at undercooling Δ *T* = 300 K for time *t* = 0.13, 0.21, 0.27 and 0.33 s. Dashed line corresponds to *F* = 1.

If one takes into account σ (*i*) dependence according to Eq. (9) with δ = 0.8 Å, after some time delay the stationary number of nuclei at ΔT = 300 K reaches at critical size: $F_{350}^S = 3.2 \times 10^{14}$ m^{-3} (experimentally reasonable value). The number of nuclei decreases with nucleus size *i* and increases with time (Figure 2). It is possible to determine maximum size i_{max} of nuclei at some preset time *t*, at which $F_{i_{max}}(t) = 1$ (dashed line in Figure 2 denotes *F* = 1) and from i_{max} (*t*) function the growth rate of nuclei can be determined.

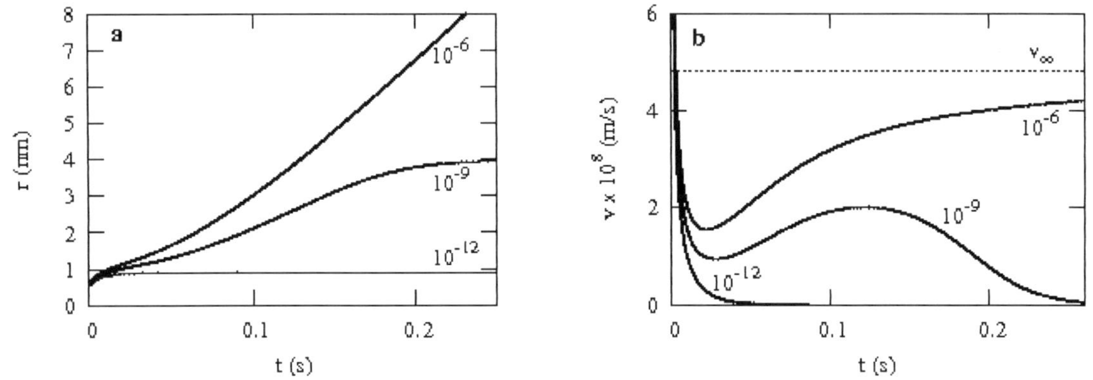

Figure 3. Radius of largest nuclei r_{max} (a) and growth rates v (b) at undercooling ΔT = 300 K in volumes $V = 10^{-6}$, 10^{-9}, and 10^{-12} m^3 as a function of time t.

Radius of the largest nuclei r_{max} increases with time (at ΔT = 300 K) for larger volumes ($V \geq 10^{-6}$ m^3). However, at smaller volumes ($V = 10^{-9}$ and 10^{-12} m^3) r_{max} reaches, at sufficiently long time, some stationary value r_{max}^S, which decreases with lowered volumes (Figure 3a). Moreover, larger nuclei are not formed in the system. The growth rates $v = dr_{max}/dt$ for a sufficiently short time reach values exceeding the growth rate within a flat interface limit v_∞ and at a certain time decreases to some minimum value, which decreases with system volume (Figure 3b). These minimum values are reached in times corresponding to the time delay of nucleation. In larger volumes ($V \geq 10^{-6}$ m^3), the growth rates increase with time close to the flat interface limit v_∞. In unit volume, size i_{max} = 194000 is attained at time t = 0.35 s and longer times would require to solve huge number of equations with increasing CPU computation time. In volume $V = 10^{-9}$ m^3 growth rate, after reaching some maximum value at t = 0.12 s, goes to zero as the nuclei of maximum size r_{max}^S are formed.

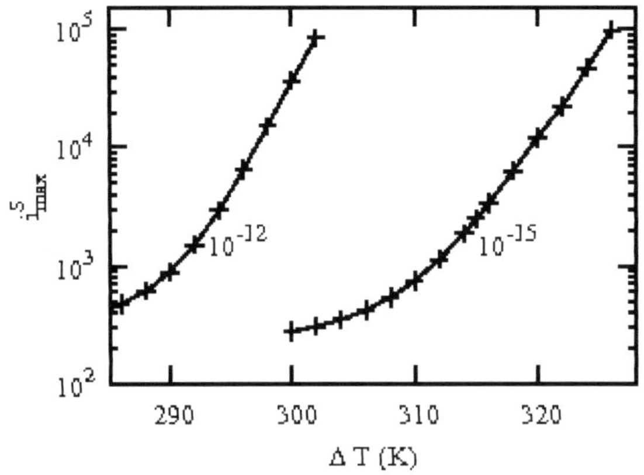

Figure 4. Stationary number of atoms in largest nuclei i_{max}^S as a function of undercooling $\Delta T = (T_E - T)$ in volumes $V = 10^{-12}$ and 10^{-15} m^3. Crosses correspond to supersaturations for which i_{max}^S was computed.

Finally, we have performed numerical solution of Eq. (1) to show how the maximum size of nuclei is influenced by undercooling near the critical value (Figure 4). At lower undercooling only subcritical nuclei are formed, *e. g.* in volume $V = 10^{-15}$ m^3 at undercooling ΔT = 300 K (critical size i^* = 350) $i_{max}^S = 280 < i^*$. It is clear that maximum size of nuclei i_{max}^S quickly increases with undercooling. At larger volume $V = 10^{-12}$ m^3 one needs lower undercooling in comparison with smaller volume $V = 10^{-15}$ m^3 to form the same maximum nucleus size.

4. Conclusion

Formation of nuclei near critical supersaturation in Ni small droplets is studied within standard nucleation model. Bokeloh et al. [7] recently presented DSC measurements and also Monte Carlo simulations to determine nucleation rate and the work of formation of nuclei in this system. We calculated the maximum size of nuclei r_{max} formed in the system from the size distribution of nuclei [condition $F_{r_{max}}(t)\ V = 1$] in various volumes V and also growth rate $v = dr_{max}/dt$ taking into account the size dependence of the interfacial energy. In small volumes a certain maximum size of nuclei r_{max}^S is reached, which increases with system volumes. The growth rate in larger volumes ($V \geq 10^{-6}$ m^3) tends to the flat interface limit with increased time. However, in small volumes, as the maximum size r_{max}^S is reached, the growth rate goes to zero via a certain maximum. The maximum number of molecules in the largest nuclei, i_{max}^S, quickly increases with undercooling and system volume. The growth of supercritical nuclei can proceed only if the number of critical nuclei is sufficiently high in nucleation system.

Acknowledgement
This work was supported by the Project no. LD15004 of the Ministry of Education, Youth and Sports of the Czech Republic (COST Action CM1402).

References

[1] Kumomi H and Shi F G 1999 Alternative Origin of Log-Normal Size Distributions of Crystallites in Controlled Solid-Phase Crystallization of Amorphous Si Thin Films *Phys. Rev. Lett.* **82** 2717–2720.
[2] Kožíšek Z, Hikosaka M, Demo P and Sveshnikov A M 2005 Nucleation kinetics of polymer formation on nucleation agent *J. Cryst. Growth* **275** e79–e83.
[3] Kožíšek Z, Hikosaka M, Okada K and Demo P 2011 Size distribution of folded chain crystal nuclei of polyethylene on active centers *J. Chem. Phys.* **134** 114904.
[4] Katz J L 1970 The Critical Supersaturations Predicted by Nucleation Theory *J. Stat. Phys.* **2** 137–146.
[5] Kožíšek Z, Sato K, Ueno S and Demo P 2011 Formation of crystal nuclei near critical supersaturation in small volumes *J. Chem. Phys.* **134** 094508.
[6] Kashchiev D 2011 Note: On the critical supersaturation for nucleation *J. Chem. Phys.* **134** 196102.
[7] Bokeloh J, Rozas R E, Horbach J, and Wilde G 2011 Nucleation Barriers for the Liquid-To-Crystal Transition in Ni: Experiment and Simulation *Phys. Rev. Lett.* **107** 145701.
[8] Kubota N 2012 Effect of sample volume on metastable zone width and induction time *J. Cryst. Growth* **345** 27–33.
[9] Toschev S, Milchev A and Stoyanov S 1972 On some probabilistic aspects of the nucleation process *J. Cryst. Growth* **13/14** 123–127.
[10] Yost F G 1974 The stochastic nature of electronucleation *J. Cryst. Growth* **23** 137–142.
[11] Uttormark M J, Zanter J W and Perepezko J H 1997 Repeated nucleation in an undercooled aluminum droplet *J. Cryst. Growth* **177** 258–264.
[12] Bokeloh J, Wilde G, Rozas R E, Benjamin R and Horbach J 2014 Nucleation barriers for the liquid-to-crystal transition in simple metals: Experiment vs. simulation *Eur. Phys. J.-Spec. Top.* **223** 511–526.
[13] Kashchiev D 2000 *Nucleation: Basic Theory with Applications* (Oxford Press, Butterworth-Heinemann Edition).
[14] Turnbull D and Fisher J C 1949 Rate of nucleation in condensed systems *J. Chem. Phys.* **17** 71-73.

[15] Merikanto J, Zapadinsky E, Lauri A and Vehkamäki H 2007 Origin of the Failure of Classical Nucleation Theory: Incorrect Description of the Smallest Clusters *Phys. Rev. Lett.* **98** 145702.

[16] Kalikmanov V I, Wolk J and Kraska T 2008 Argon nucleation: Bringing together theory, simulations, and experiment *J. Chem. Phys.* **128** 124506.

[17] Tolman R C 1949 The effect of droplet size on surface tension *J. Chem. Phys.* **17** 333–337.

[18] Kožíšek Z and Demo P 2017 Homogeneous crystal nucleation in Ni droplets *J. Cryst. Growth* **475** 247–250.

Hydrogen treatment of titanium based alloys

M Losertová, M Hartmann, I Schindler and J Drápala

VŠB – Technical University of Ostrava, Faculty of Metallurgy and Materials Engineering, 17. listopadu 15, 708 33 Ostrava, Czech Republic

mlosertova@vsb.cz

Abstract. The positive effect of the hydrogen on hot deformation behaviour at 700 and 750 °C was investigated after thermal hydrogen treatment of Ti6Al4V and Ti26Nb alloys. Comparing the results obtained for the non-hydrogenated and hydrogenated specimens of both alloys, it was found that the hydrogen content as high as 1325 wt. ppm has an obvious benefit effect on high temperature deformation behaviour in the Ti6Al4V alloy by stabilizing beta phase and lowering thermal deformation resistance. In the case of Ti26Nb alloy the hydrogen content of 2572 wt. ppm suppressed stress instabilities during hot compression but slightly increased thermal deformation resistance. The microstructure study was performed before and after the isothermal compression tests on the specimens in hydrogenated as well as in non-hydrogenated condition. The hydrogen amounts in the specimens were measured by means of an analyser LECO RH600.

Keywords: hydrogen effect, hydrogen embrittlement, hydrogen induced plasticity, biocompatible alloys.

1. Introduction

Titanium based materials are frequently used in surgery and orthopedics. Among them Ti6Al4V alloy belongs to the most applied alloys. As compared with 316L stainless steel, Ti6Al4V is characterized by high strength-to-weight ratio, good corrosion resistance, higher biocompatibility and lower Young's modulus [1-4]. Nevertheless, workability of this alloy at room temperature is difficult and forming must be performed at elevated temperatures, thus improving hot deformation behavior by decreasing flow stress and working temperature is desirable. The essential reason of difficulties with forming consists in the microstructure formed of $\alpha + \beta$ phases and in the relatively high temperature of $\alpha + \beta \rightarrow \beta$ transformation whereas β phase with body centered cubic structure is more malleable then α phase with hexagonal close packed lattice. The transition temperature is called β transus and for Ti6Al4V reaches 980 °C.

Although hydrogen deteriorates mechanical properties of many engineering metallic materials, titanium alloys included, and the brittleness of conventional alloys induced by the hydrogen is observed at room temperature as well as at elevated temperatures, in case of Ti alloys the hydrogen can act as a temporary alloying element in special heat treatment technology, so-called the thermo-hydrogen treatment (THT) [5-7] and shows beneficial effect in improving particular properties. Whereas the hydrogen works as β stabilizer in Ti alloys, THT technology employs the reversible reaction of hydrogen with titanium in order to modify phase compositions, kinetics of phase transformations and evolution of metastable phases in titanium alloys [8-11], that provides decreasing

Content from this work may be used under the terms of the Creative Commons Attribution 3.0 licence. Any further distribution of this work must maintain attribution to the author(s) and the title of the work, journal citation and DOI.
Published under licence by IOP Publishing Ltd

flow stress and increasing hot workability of titanium alloys at lower temperatures. The temporary hydrogen alloying allows the microstructure refinement resulting in enhancement of the fatigue and strength properties [12].

Recently new candidates promising for fixation devices through the decreased Young's modulus approaching that of human bone are β alloys (or more precisely called as metastable β alloys). The alloy compositions based on non-toxic elements, as Nb, Mo, Ta and Zr, enable to reach higher biocompatibility comparing to Ti6Al4V and the other commonly used metallic alloys [1,4]. Although the refractory alloying elements mentioned above act as β stabilizers, the microstructure of β alloys may contain some α phase precipitates depending on amount of β alloying additions. Even if the β alloys are better workable, remaining α phase can cause some problems during forming processes.

The present work presents the study of hydrogen effect on hot deformation behavior at two different temperatures and microstructure evolution after the isothermal compression test performed for Ti6Al4V and Ti26Nb alloys.

2. Experimental

To investigate the hydrogen effect on the microstructure and hot deformation behavior of the non-hydrogenated and hydrogenated Ti based alloys specimens, isothermal uniaxial compression test (CT) at 700 and 750 °C was performed on Gleeble-3800 machine using basic CTA module. The experimental specimens were prepared from the forged bars of Ti6Al4V (in wt.%) alloy supplied by company MEDIN, a. s. and of Ti26Nb (in at.%) (Ti40.5Nb in wt.%) alloy prepared in own laboratories. The cylindrical samples of 10 mm in diameter and 12 mm in height exactly were machined and treated by three steps thermal treatment. In order to compare the high temperature deformation behavior with and without hydrogen in the alloy structure the same procedure of thermal treatment (TT), according to [10,13] and described in Figure 1, was performed for one group of the cylinders in flowing argon (purity 5N) and for second group in flowing hydrogen (purity 5N). Considering higher solubility of the hydrogen predominantly in β phased TiNb [11], the times of hydrogenation steps were modified [14].

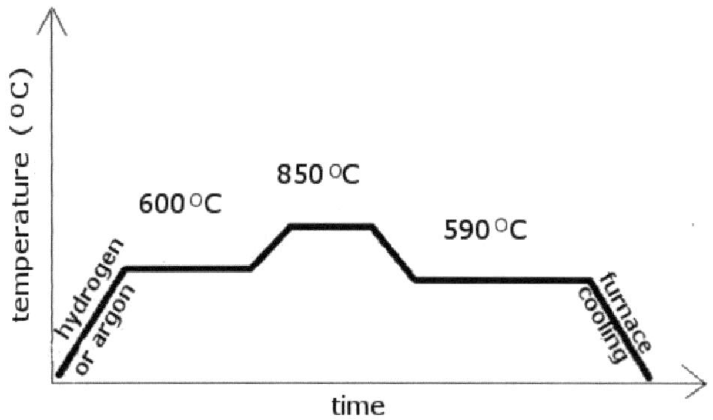

Figure 1. Three steps regime of thermal treatment with modified times at holding temperatures for both alloys (Ti6Al4V and Ti26Nb) before subsequent isothermal compression test.

After heating with the rate of 1200 $°C \cdot h^{-1}$ to 600 °C in hydrogen (or in argon) the temperature was kept for 60 and 30 minutes for Ti6Al4V and Ti26Nb, respectively. Then the temperature was increased to 850 °C and held for 30 and 18 minutes for Ti6Al4V and Ti26Nb, respectively. After that the cooling to 590 °C and alloy aging was performed for 240 and 60 minutes for Ti6Al4V and Ti26Nb, respectively. Finally, the cooling was realized in the furnace to 400 and 300 °C in hydrogen (or in argon) for Ti6Al4V and Ti26Nb, respectively, and accomplished in argon to the room temperature for both alloys.

During hot deformation test, the specimens were heated with a heating rate of 5 °C·h^{-1} and soaked for 3.0 min at the deformation temperature before the isothermal compression. The specimens of both alloys were deformed to 50 % of their initial height with a constant initial strain rate of 0.005 s^{-1} and then air-cooled to room temperature. The flow stress during hot compression was recorded as a function of the strain. The test was stopped at the strain of 0.7.

The microstructure evolution of both alloys in different conditions of thermo-mechanical treatment was observed using optical microscopy. All metallographic specimens were ground and polished in transversal and longitudinal sections by conventional methods and then etched in solution composed of 2 HF : 4 HNO_3 : 80 H_2O. The amount of hydrogen in the specimens was measured by means of an hydrogen determinator LECO RH 600.

3. Results and discussion

The hot deformation behavior at 700 and 750 °C during the compression test (CT) of non-hydrogenated and hydrogenated cylinders of Ti6Al4V and Ti26Nb are presented in Figure 2. Comparing the macroscopic views of hydrogenated hot compressed specimens of both alloys that are shown in Figure 3, it is clear that despite the different alloy compositions and high hydrogen contents in THT treated cylinders the hot deformation behavior is macroscopically similar. Even after the 50 % reduction in height that resulted in barrel-shaped samples, no cracks were observed on the surfaces of Ti6Al4V (Figure 3a). Unlike Ti26Nb, the presence of hydrogen in the structure decreased the peak of flow stress about by 50 and 100 MPa at 700 and 750 °C, respectively. All samples exhibited dynamic recrystallization that decreased flow stress and helped to better formability. The true stress - true strain curves indicate that the mechanisms of softening are sufficiently fast to balance the rate of work hardening and are suggestive of mechanisms like dynamic recrystallization, superplasticity or dynamic recovery occurring at very high rates.

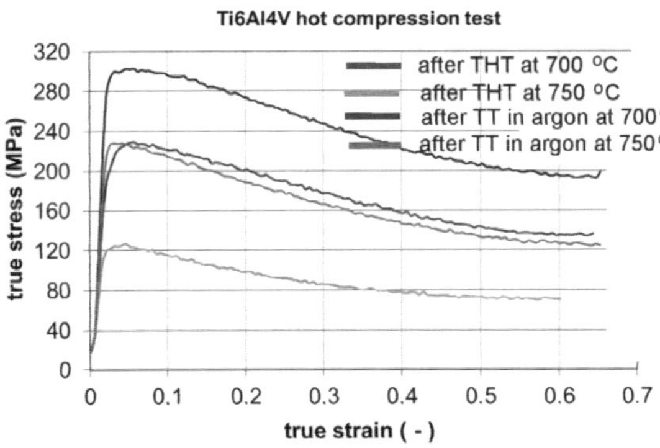

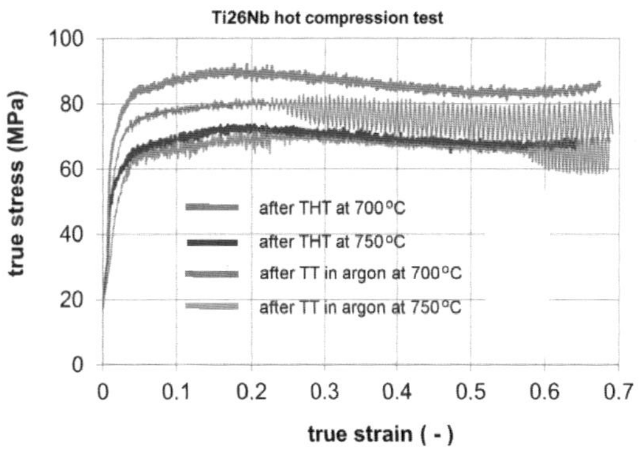

Figure 2. Comparison of CT curves at 700 and 750 °C and strain rate 5·10^{-3} s^{-1} after thermal treatment in argon and hydrogen for a) Ti6Al4V and b) Ti26Nb alloys.

Figure 3. Specimens after the THT and compression test of a) Ti6Al4V and b) Ti26Nb (cracking in red boxes).

The microstructure of the as-received Ti6Al4V alloy was formed of bimodal equiaxed grains of α and (α+ β). After thermal treatment in Ar, the similar microstructure was observed indicating that annealing time did not affect grain growth or recrystallization (Figure 4a). The microstructure of thermo-hydrogen treated (THT) specimens was markedly changed as it can be seen in Figure 4b. During holding at 590 °C the microstructure fully decomposed after following eutectoid reaction: $\beta_H \rightarrow \alpha + \beta_H \rightarrow \beta_H + (\alpha + \gamma)$, according to [8-11], where γ is titanium hydride. The deformation microstructure after the isothermal compressive test quite differs for both testing temperatures, as seen in Figure 4. Retained amount of hydrogen after isothermal compression test resulted in the martensite microstructure with rare acicular α grains that indicated the transformation of β phase to two phase α + β microstructure processed at lower temperatures then it would be currently for hydrogen free-alloys. No hydrides were observed in all Ti6Al4V microstructures by optical microscopy.

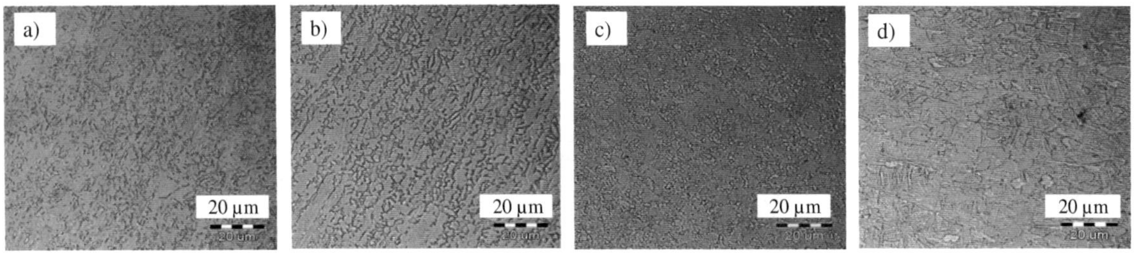

Figure 4. Microstructure of Ti6Al4V a) after TT in Ar without formability test; b) after THT without formability test; c) after TT in Ar and compressed at 750 °C and strain rate 0.005 s^{-1}; d) after THT and compressed at 750 °C and strain rate 0.005 s^{-1}

Figure 2a shows the true stress - true strain curves of Ti6Al4V specimens at 700 and 750 °C in the both thermal treatment conditions. In the case of Ti6Al4V the stress is increasing with raising strain to a maximum and then goes down, which shows that the working softening has prevailed. The deformation evolution of THT specimen shows that softening effect from dynamic recovery passed to equilibrium with the work hardening and then passed into a quasi-steady flow stage, during which the stress decreasing slowed. So compared with the non-hydrogenated state, it can be seen that hydrogen didn't delay the dynamic recovery of the deformation in titanium alloy, conversely, the steady flow stress for hydrogen containing alloy was much lower than that for non-hydrogenated specimen at the same compression temperature. This result proved that at as high concentrations of hydrogen as 1325 wt. ppm the flow stress was obviously reduced and the cylindrical sample was deformed without cracking.

The microstructure of non-hydrogenated Ti26Nb samples was formed of β grains with fine α precipitates (Figure 5a). After the hot compression tests, the microstructures of alloy were dynamically recrystallized to large β grains with very fine subgrains (Figures 5b and 5c). The hydrogen charging led to the important refining of microstructure after the THT as well as after the CT, as seen in Figure 6a. No hydrides were observed by optical microscopy.

The macroscopic view of the Ti26Nb specimens in THT condition and after the compression tests with an amount of reduction in height of 50 % is illustrated in Figure 3b. The surfaces of hydrogenated specimens were cracked (red boxes in figure) while the non-hydrogenated specimens were cracked rather rarely on the surface.

The flow stresses recorded for the Ti26Nb alloy are depending on the temperature. It is clear that niobium is β stabilizer, so the flow stresses are lower for the Ti26Nb than for Ti6Al4V due to the higher fraction of more ductile β phase. Comparing the deformation behavior at two different temperatures for hydrogen non-charged and charged specimens, it is evident that with increasing temperature the flow stress is decreasing and by reason of this the hot forming is improved for both thermal treated stages. In the case of the hydrogen non-charged specimens unusual behavior was registered during the CT (Figure 2b). Large instabilities of flow stress were detected that can be related with lower rate of dynamic recrystallization.

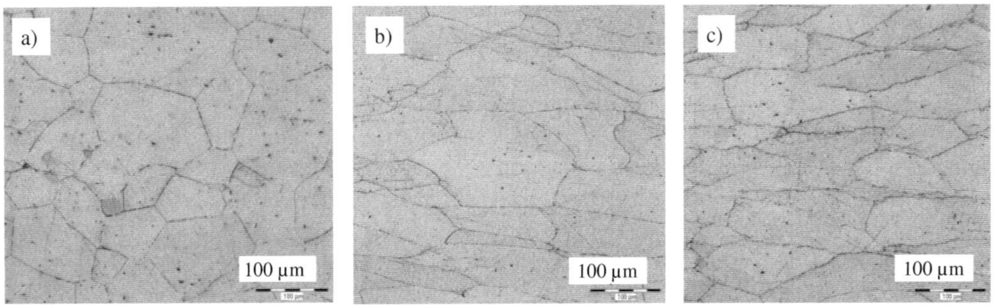

Figure 5. Optical micrograph of Ti26Nb alloy treated by three steps TT in Ar a) without compression test; b) compressed at 700 °C and c) compressed at 750 °C.

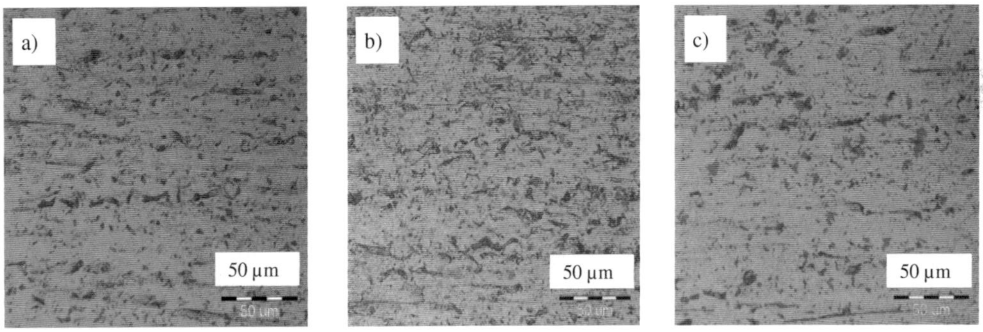

Figure 6. Optical micrograph of Ti26Nb alloy a) treated by three steps THT and deformed by compression test at b) 700 °C and c) 750 °C.

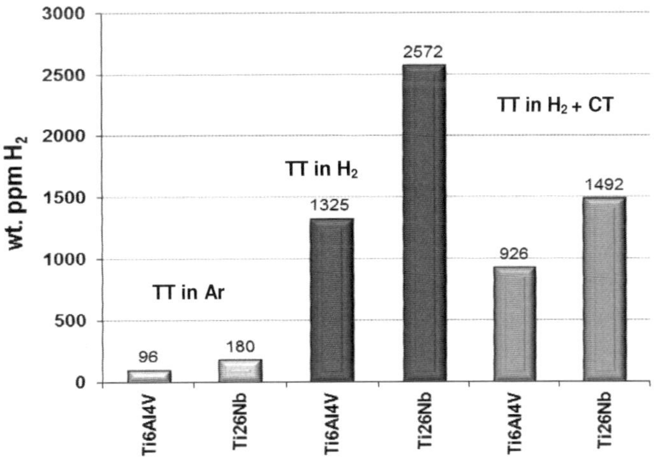

Figure 7. Amount of hydrogen in the tested specimens.

The hydrogen contents measured for four conditions of thermo-mechanical treatment (TT in argon or hydrogen, TT in argon or hydrogen + compression tested) of both Ti6Al4V and Ti26Nb alloys are summarized in Figure 7. The highest amount (2572 wt. ppm) was determined in Ti26Nb after the THT that corresponds with higher hydrogen solubility in β phase. After the compression test, the content of hydrogen in cylinders decreased from 1325 to 926 wt. ppm and from 2572 to 1492 wt. ppm for Ti6Al4V and Ti26Nb, respectively. Nevertheless, the hydrogen concentration was sufficient for retaining β phase. The hydrogen amount decrease after the CT was more important just for the alloy with niobium (by 42 % for Ti26Nb and 31 % for Ti6Al4V). Unfortunately, the hydrogen promoted cracking on the surfaces of compression tested Ti26Nb specimens. Unlike the Ti6Al4V alloy, the hydrogen content in the Ti26Nb specimens increased the flow stress but on the other side positively influenced the flow stress instabilities observed on the true stress - true strain curves of non-hydrogenated stages (Figure 2b). Indeed, the evolution of uniaxial compressive test curves showed the stabilization of flow stress due to the hydrogen for high strains at both temperatures. The dissimilarity in hot compression behavior of both Ti alloys for hydrogen charged and non-charged stages are related with high niobium content and therefore β metastable microstructure, where hydrogen is effecting strengthening by solid solution [15]. The hydrogen effect on the development of the compression test was inconsistent for both alloy compositions.

4. Conclusion

The hydrogen effect on the hot compression behavior was investigated for two Ti alloys. The hot deformation behavior and microstructure evolution of Ti6Al4V and Ti26Nb alloys were investigated after three steps thermal treatment in flowing hydrogen or argon gases. Comparing the results obtained on the base of the isothermal uniaxial compression tests at 700 and 750 °C for non-hydrogenated and hydrogenated specimens, it can be concluded following:

- The hydrogen content has an obvious benefit effect on high temperature deformation behavior of the Ti6Al4V alloy by decreasing the β transus temperature and lowering the hot deformation resistance. Retained amount of hydrogen after isothermal compression test resulted in the martensite microstructure with rare acicular α grains that indicates the transformation of β phase at temperatures at which the hydrogen free-alloys are formed normally of two phase α + β microstructure.
- It was found that the hydrogen content as high as 1325 wt. ppm for Ti6Al4V significantly suppressed the flow stress for both temperatures of hot compression without inducing the sample cracking.
- The hydrogen content as high as 2572 wt. ppm for Ti26Nb proved an obvious positive effect on the suppression of the stress instabilities for both temperatures.
- However, the high amount of hydrogen induced increasing flow stress and resulted in cracking of the specimens.
- The benefit effect of hydrogen on increasing formability of Ti26Nb alloy at lower temperatures was not evident as it was observed in the case of Ti6Al4V alloy.
- The hydrogen amounts measured after the different thermal treatment were evaluated and the highest content (2572 wt. ppm) was determined in Ti26Nb after the THT that corresponds with higher hydrogen solubility in β phase. After the compression test the hydrogen content decreased by 42 % that was more important than in the case of Ti6Al4V alloy.

Acknowledgment

This article has been elaborated in the framework of the project No. LO1203 "Regional Materials Science and Technology Centre - Feasibility Program" funded by Ministry of Education, Youth and Sports of the Czech Republic and co-financed by the European Social Fund and of the projects SGS SP2017/77 and SGS SP2017/58 supported by Ministry of Education, Youth and Sports of the Czech Republic.

References

[1] Niinomi M 2008 Mechanical biocompatibilities of titanium alloys for biomedical applications *J. of the mechanical behavior of biomedical materials* **1** 30-42

[2] Enderle J D, Bronzino J D and Blanchard S M 2005 *Introduction to Biomedical Engineering* (Second ed.) Elsevier Academic Press Amsterdam 1253 p ISBN 978-0123749796

[3] Sloten J V, Labey L, Van Audekercke R and Van der Perre G 1998 Materials selection and design for orthopedic implants with improved long-term performance *Biomaterials* **19** 1455–1459

[4] Geetha M, Singh A K and Asokamani R 2009 Ti based biomaterials, the ultimate choice for orthopaedic implants – A review *Progress in Materials Science* **54** 397-425

[5] Eliaz N, Eliezer D and Olson D L 2000 Hydrogen-assisted processing of materials *Mat. Sci. Eng. A* **289** 41-53

[6] Nelson H G 2013 in: A W Thompson and N R Moody (Eds) *Hydrogen Effects in Materials* Wiley Online Library p 699 DOI: 10.1002/9781118803363

[7] Senkov O N and Froes F H 1999 Thermohydrogen processing of titanium alloys *Int. J. Hydrogen Energy* **24** 565-576 DOI: 10.1016/S0360-3199(98)00112-8

[8] Qazi J I, Senkov O N, Rahim J, Genc A and Froes F H (Sam) 2001 Phase Transformations in Ti-6Al-4V-xH Alloys *Metall. Mater. Trans. A* **32** 2453-2463

[9] Murzinova M A, Mazurski M I, Salishev G A and Afonichev D D 1997 Application of reversible hydrogen alloying for formation of submicrocrystalline structure in (alpha + beta) titanium alloys *Proc. of the Selected Papers of the First Int. Conf. on Hydrogen Treatment of Materials* Int. Assoc. for Hydrogen Energy **22** 201–204

[10] Fang T Y and Wang W H 1998 Microstructural features of thermo mechanical processing in a Ti-6Al-4V alloy *Materials Chemistry and Physics* **56** 35-47 PII S 0254-0584(98)00137-0

[11] *Phase Diagrams of Binary Hydrogen Alloys* 2000 Ed. by F D Manchester ASM International pp 238-258

[12] Nalla R K, Boyce B L, Campbell J P, Peters J O and Ritchie R O 2002 Influence of microstructure on high cycle fatigue of Ti–6Al–4V: bimodal vs. lamellar structure *Metall. Mater. Trans. A* **33** 899–918

[13] Losertová M, Štamborská M and Schindler I 2014 Influence of thermo hydrogen treatment on hot deformation behavior and microstructural evolution of Ti-6Al-4V alloy *Archives of Mat. Sci. Eng.* **66** 62-66

[14] Losertová M, Štefek O, Schindler I, Szkandera L and Kudela P 2017 Hydrogen effect on compression mechanical properties of TiNb alloys at elevated temperatures *IOP Conf. Series: Mater. Sci. Eng.* **179** 012048 doi:10.1088/1757-899X/179/1/012048

[15] Zong Y Y, Huang S, Wang Y and Shan D B 2014 Strenghening versus softening mechanisms by hydrogen addition in beta-Ti40 alloy *Int. J. Hydrogen Energy* **39** 3498-3504

Microstructure evolution of heat treated NiTi alloys

M Losertová[1], M Štencek[1], D Matýsek[2], O Štefek[1] and J Drápala[1]

[1] VŠB – Technical University of Ostrava, Faculty of Metallurgy and Materials Engineering, 17. listopadu 15, 708 33 Ostrava-Poruba, Czech Republic
[2] VŠB – Technical University of Ostrava, Faculty of Mining and Geology, 17. listopadu 15, 708 33 Ostrava-Poruba, Czech Republic

E-mail: mlosertova@vsb.cz

Abstract. Superelastic behavior of off-stoichiometric NiTi alloys is significantly affected by microstructure changes due to heat treatment. Applying appropriate thermal treatments important effects on microstructural changes, transformation temperatures and thermomechanical properties of final NiTi products can be achieved. The experimental samples of NiTi alloy with 55.8 wt.% Ni were submitted to heat treatment and the microstructures before and after the treatment were observed. The thermal regimes consisted of annealing treatment at 600 °C for 1 hour followed by water quenching and of ageing at eight different temperatures (250, 270, 290, 300, 350, 400, 450 and 500 °C) for 30 minutes. Microstructure features studied by means of optical and scanning electron microscopies, EDX microanalyses, X-ray diffraction analyses and microhardness measurement, have shown that higher ageing temperatures led to microstructure changes and corresponding increase in microhardness.

Keywords: NiTi, annealing, ageing, XRD, martensite, Ni_4Ti_3.

1. Introduction

The intermetallic NiTi-based alloy, usually called Nitinol, belongs to a group of materials known as Shape Memory Alloys (SMA); sometimes also called as "Smart Materials". Its chemical composition ranges between 49 and 51 at.% Ni. Beside shape memory behavior and superelasticity, very high corrosion resistance and good biocompatibility due to TiO_2 surface layer should be mentioned.

The control of functional properties of NiTi alloys only through chemical composition changes is insufficient in many cases and is not able to solve various problems for particular applications. Heat treatment is a different, and in many regards preferable, alternative, which influences significantly a lot of resulting properties, such as a microstructure, a level of internal stresses, a homogenous distribution of alloying elements, a character of segregated phases or precipitates and many other characteristics. In a case of NiTi alloy, heat treatment has also a crucial effect on the shape memory behavior and superelasticity, therefore this issue is necessary to be dealt closely. However, a general summary of a heat treatment effect on NiTi alloys is very complicated and strongly depends on a particular material composition. Generally, with a higher Ni concentration in NiTi alloys, precipitation processes occur during thermal or thermo-mechanical treatments unlike Ti-rich compositions. Ageing treatment promotes the formation of various Ni-rich precipitates, such as Ni_4Ti_3, Ni_3Ti_2 or Ni_3Ti and the occurrence of multi-stage transformation behavior including the R phase transformation associated with the formation of fine and coherent Ni_4Ti_3 particles [1-4]. As the most commercial NiTi alloys

Content from this work may be used under the terms of the Creative Commons Attribution 3.0 licence. Any further distribution of this work must maintain attribution to the author(s) and the title of the work, journal citation and DOI.
Published under licence by IOP Publishing Ltd

belongs to Ni-rich alloys, their transformation and mechanical properties can be affected by developing Ni_4Ti_3 precipitates.

It is known, that the temperature of reverse phase transformation austenite ↔ martensite can be changed significantly through modifications of a heat treatment temperature, a time at the temperature or a cooling rate and a cooling method. Generally, increasing ageing temperature within range from 400 to 600 °C decreases the temperature of reverse transformation [3-6]. Thus, applying various annealing and ageing treatments important differences in structural phases and transformation behavior can be achieved in final products [6].

The work was focused on the study of microstructure changes in NiTi alloy occurring after application of thermal treatment at different temperatures.

2. Experimental

The experimental binary NiTi alloy was supplied by MEDIN, a.s. in form of a wire with a diameter of 1 mm. The chemical composition of the alloy corresponded to Ni content of 55.8 wt.% (which is equivalent to 50.7 at.%). Ten specimens of lengths within the range from 1.5 to 2 cm were taken from the as-received NiTi wire. One specimen was maintained in the as-received condition as reference sample, the others were heat treated in Linn HT-1800 chamber furnace with resistance heating. The heat treatment under argon atmosphere involved annealing at 600 °C for 1 hour with following water quenching and ageing at eight temperatures (250, 270, 290, 300, 350, 400, 450 and 500 °C) for 30 minutes with following water quenching, as seen in Figure 1.

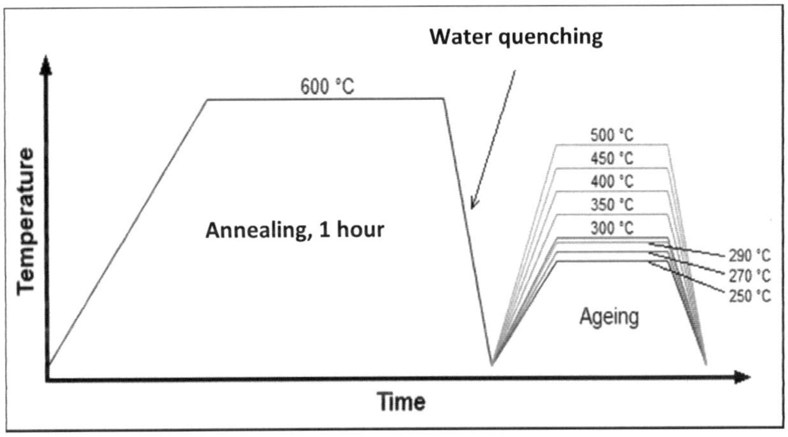

Figure 1. Schematic of heat treatment of NiTi samples.

The microstructure was studied on metallographically prepared and chemically etched (solution of HF + HNO_3 + CH_3COOH) specimens using the optical microscopy on OLYMPUS GX51. The phase analysis was performed by means of scanning electron microscope (SEM) JEOL JSM-6490LV equipped with INCA x-act analyzer. The Vickers microhardness was measured by FUTURE-TECH FM-100 automatic microhardness tester with FM-ARS900 control unit. The microhardness values were determined by twelve indents in central line parallel with the sample edge at load of 0.1 kg.

To determine the phases present in the NiTi alloy after different heat treatments X-ray diffraction (XRD) was performed using a Bruker-AXS D8 Advance system with 2θ/θ geometry of measure operated at 40 kV and 40 mA and equipped with Lynx-Eye PSD (CuKα source/Ni filter, step size of 0.014° 2θ, dwelling time of 46 s per step). Data refinement, fitting and background elimination was done using Diffrac.EVA, version 2 software and PDF-2 database or COD respectively. Phase identification by X-ray diffraction was performed on the surface of 0.6 x 12 mm approximately at 45 ° 2θ and verified by Rietveld method modeling using Bruker Topas, version 5. Input data were adapted of COD database. Martensite phases presence on XRD spectra were verified using Le Bail method modeling where input data were adapted of PDF database.

3. Results and discussion

The microstructure of selected specimens in the as-received stage, after annealing at 600 °C for 1hour and after ageing at various temperatures are presented in Figures 2 - 6. The SEM observation and EDX microanalyses of NiTi specimens in different heat treatment conditions showed that the matrix consists of a NiTi near-equiatomic phase, in which a minority $NiTi_2$ phase occurs in a form of broken particle clusters elongated in a direction along longitudinal axis of the wire, i.e. along the wire drawing direction (Figures 2, 3a and 5a). The specimens in as-received stage showed martensitic structure that was markedly changed after annealing at 600 °C. Ageing at temperatures of 250, 270, 290, 300, 350, 400, 450 and 500 °C resulted in precipitation of a high number of fine disk-shaped precipitates, unlike martensite needles and plates that with increasing annealing temperature seemed to coarsen.

On the base of the X-ray diffraction analysis (Figure 7) it is evident that the phase compositions of the specimens are complex and formed of structure similar phases. The XRD spectra contain more intensive and strongly superimposed intensities only in relatively narrow range from 38.5 to 45.5° 2θ CuKα. The intensities are evidently extended that indicates limited size of diffraction areas. Although the phase compositions are very similar in all specimens, the differences in the ratio of intensities corresponding to martensite and austenite phases were observed.

The precipitates Ni_4Ti_3 and $NiTi_2$ were identified as minority phases. Several martensite types were revealed, it means martensites with space group P21/m and P21/c, as well as R phase NiTi. The as-received sample is formed of evidently monoclinic super-textures P21/m + P2/c and cubic NiTi that is evidently faulted or highly textured for (003) diffraction being is quite missing. According to XRD analysis and metallographic observation, the structure is formed of ledge-like martensite aggregates, in which the submicron cubic parent NiTi is overgrown. Relatively high abundance of austenite and high crystallinity is observed in the sample aged at 400 °C. The annealing at 600 °C led to lowest content of austenite phase.

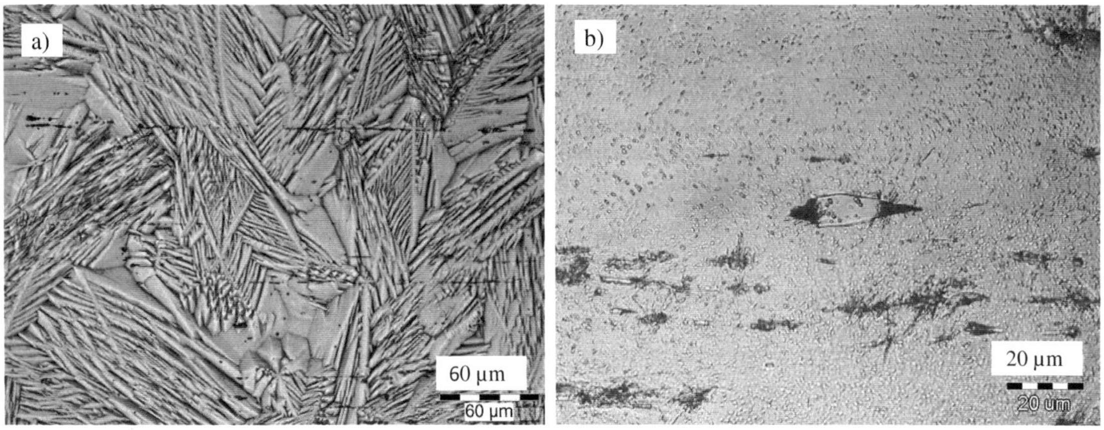

Figure 2. Microstructure of NiTi samples a) as-received and b) annealed at 600 °C for 1 h and water quenched.

As it was mentioned above two precipitates types were observed in metallographic observation and proved by XRD spectra. Whereas the Ni_4Ti_3 precipitation processes can be observed mainly in Ni-rich alloys containing more than 50 at.% Ni [5], the disc-shaped precipitates revealed in our case using SEM (Figures 5b and 6) are of Ni_4Ti_3 type. Their length ranged approximately between 0.7-1.3 μm, the thickness within an interval of 0.1-0.25 μm.

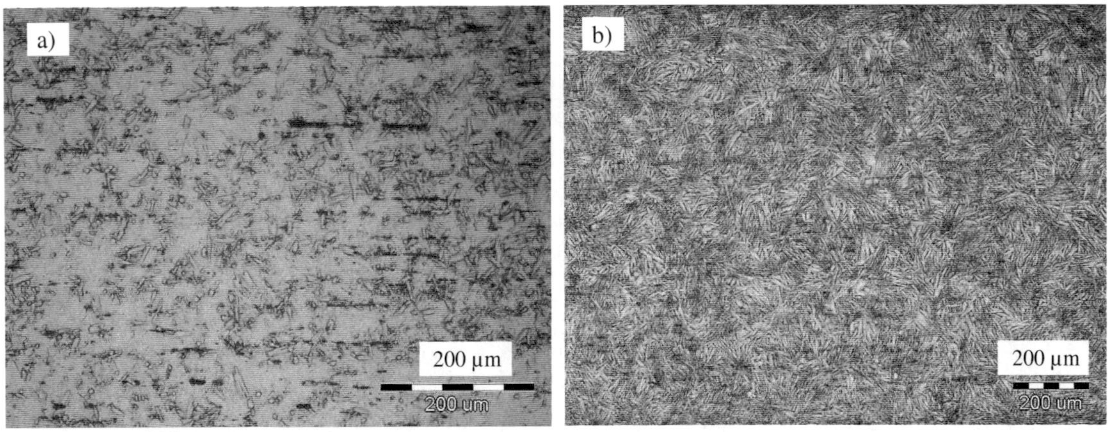

Figure 3. Microstructure of NiTi samples annealed at 600 °C for 1 h, water quenched and aged for 30 minutes a) at 290 °C and b) at 350 °C.

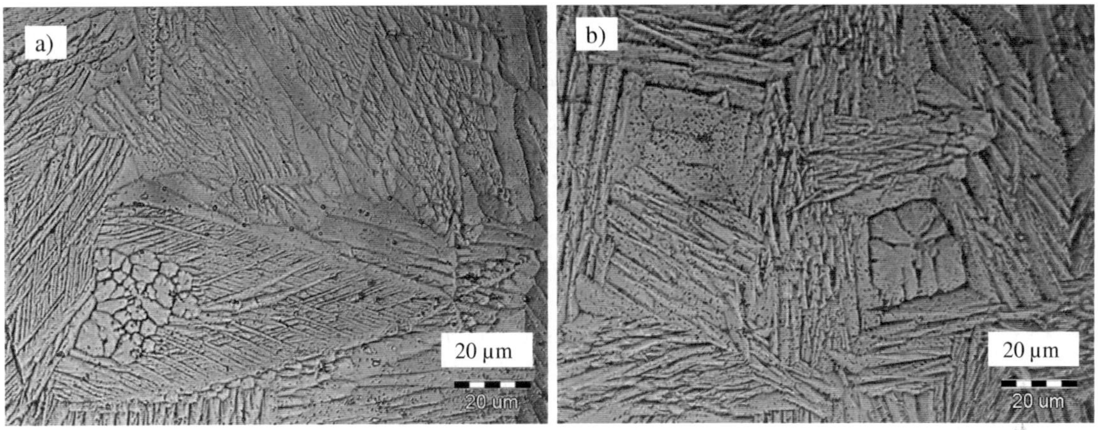

Figure 4. Microstructure of NiTi samples annealed at 600 °C for 1 h, water quenched and aged for 30 minutes a) at 400 °C and b) at 450 °C.

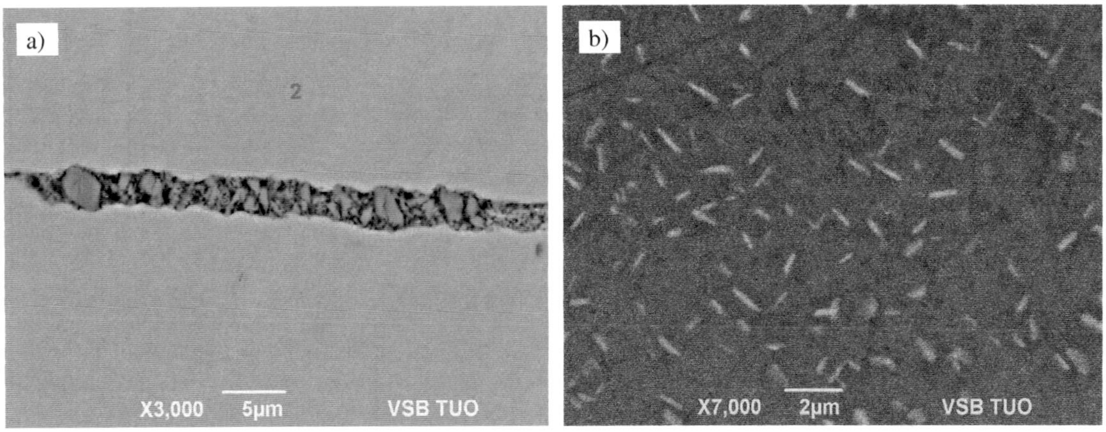

Figure 5. SEM microstructure of NiTi samples a) as-received (red indexes: 1 - $NiTi_2$, 2 - NiTi matrix, 3 - fractured $NiTi_2$); b) after annealing at 600 °C for 1 h and water quenching.

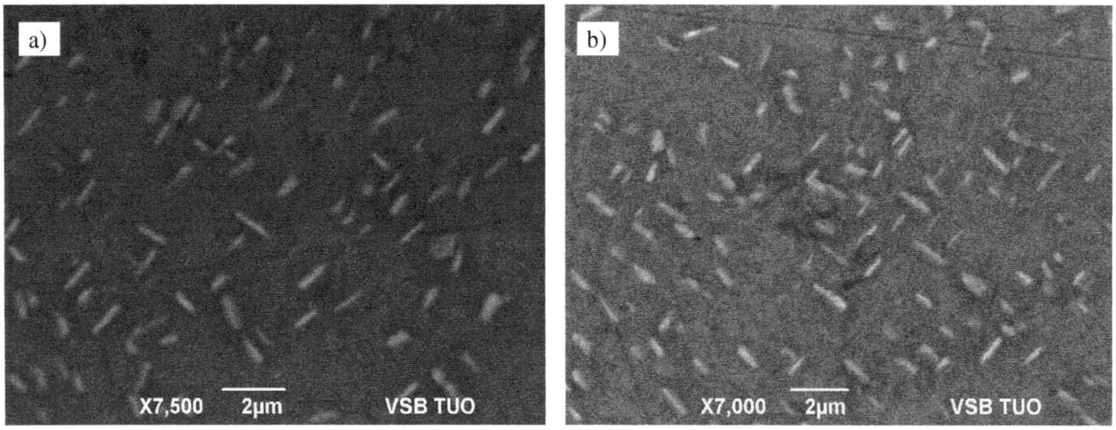

Figure 6. SEM microstructure after annealing at 600 °C for 1 h, quenching and ageing for 30 minutes a) at 290 °C and b) at 300 °C.

Owing to their very small size the determination of chemical composition by EDX analyses was not possible to perform. The morphology and the structure properties of very fine Ni_4Ti_3 precipitates were confirmed in [6].

Unlike the as-received NiTi wire with martensitic microstructure, at 600 °C heat treated specimen developed the transformation into austenite structure that after water quenching passed to multiphase composition in accordance with XRD analyses in the Figure 7b. Martensite needles were very fine and only exceptionally in the neighborhood of the $NiTi_2$ particles reached the well detectable sizes, as seen in the Figure 2b.

An influence of ageing on a change of the $NiTi_2$ phase particles was not observed. However, with an increasing temperature of ageing a coarsening of the martensitic structure occurred (Figure 3). The increasing temperature gradient (along with the increasing temperature of ageing), which was then a driving force for the martensitic transformations during the following quenching of the specimens, can be considered the main factor enabling an occurrence and development of martensite.

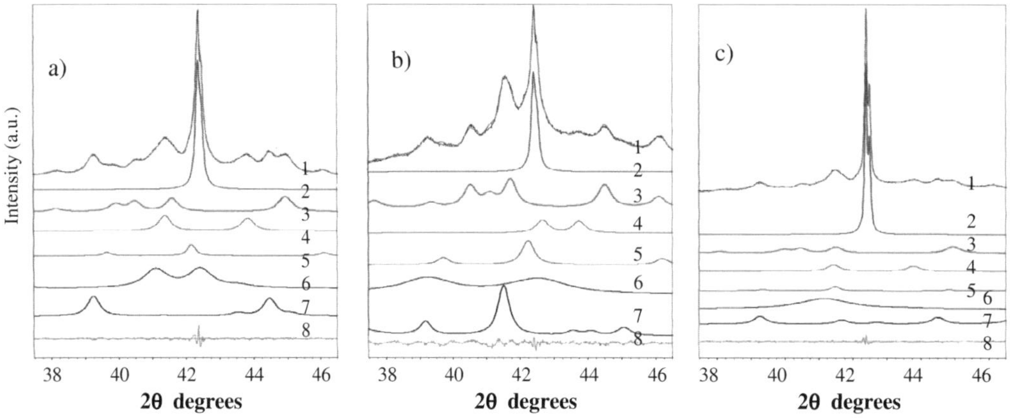

1 – measured and calculated model, 2 – NiTi austenite, 3 – P21/m martensite (B19), 4 – Ni_4Ti_3 hexagonal phase, 5 – $NiTi_2$ cubic phase, 6 – NiTi R phase, 7 – P21/c martensite (B19′), 8 – difference profile model

Figure 7. XRD charts for a) as-received wire, b) annealed at 600 °C and c) annealed at 600 °C and aged at 400 °C.

The average results of HV0.1 microhardness are summarized in Table 1. Comparing the as-received specimen and the specimens after annealing at 600 °C, an appreciable drop in HV0.1 microhardness was caused by the transformations from coarse martensite structure into fine martensite one. Afterwards a well evident increase in the microhardness values was noted with increasing ageing temperatures that corresponded to changes in the martensite and austenite proportion in the structure as well as to the Ni_4Ti_3 precipitation.

Table 1. Microhardness HV0.1 of NiTi samples after various conditions of heat treatments.

Heat treatment	As-received	Annealed (°C)	Temperature (°C) of ageing for 30 minutes							
		600	250	270	290	300	350	400	450	500
Microhardness HV0.1	357 ± 19	246 ± 18	253 ± 17	291 ± 19	301 ± 22	349 ± 36	323 ± 23	315 ± 18	320 ± 25	320 ± 14

The above mentioned results imply that the phase composition is critical parameter for requirements to influence the superelastic behavior of NiTi applications. It is clear, that the choice of a particular heat treatment affects strongly above all the transformation temperature values (A_s, A_f, M_s and M_f) and precipitation processes.

5. Conclusion

Based on the study of microstructure changes in NiTi alloy occurring after thermal treatment at different temperatures several conclusions can be drawn:

1. It was proved by means the optical microscopy, SEM/EDX and XRD methods that applying heat treatment (annealing at 600 °C and following ageing) can affect precipitation processes in the NiTi alloy in favor of very small Ni_4Ti_3 particles.
2. No essential changes in the morphology and size of $NiTi_2$ precipitates in the matrix were observed.
3. The tiny disk-shaped morphology implies that these ones are of Ni_4Ti_3 type even the EDX analyses couldn't confirm the stoichiometric composition of the particles.
4. The following ageing regimes led to the change in proportional abundance of martensite/austenite.
5. The decrease in microhardness values after annealing at 600 °C and their slow increasing with growing ageing temperature attached the changes in microstructure.

Acknowledgement
This article has been elaborated in the framework of the project No. LO1203 "Regional Materials Science and Technology Centre - Feasibility Program" funded by Ministry of Education, Youth and Sports of the Czech Republic and co-financed by the European Social Fund, of the project No. TH01020487 "Development of endodontic tools" financed by Technology Agency of the Czech Republic and of the projects SGS SP2017/77 supported by Ministry of Education, Youth and Sports of the Czech Republic.

References

[1] Jiang F, Liu Y, Yang H, Li L and Zheng Y 2009 Effect of aging treatment on the deformation behaviour *Acta Materialia* **57** 4773–4781
[2] Srivastava A K, Yang Z, Schryvers D and Van Humbeeck J 2008 Effect of annealing on cold-rolled Ni–Ti alloys *Materials Science and Engineering:* A **481-482** 594-597

[3] Zheng Y, Jiang F, Li L, Yang H and Liu Y 2008 Effect of ageing treatment on the transformation behaviour of Ti–50.9 at.% Ni alloy *Acta Mater* **56** 736-745
[4] Bhagyaraj J, Ramaiah K V, Saikrishna, C N, Bhaumik S K and Gouthama 2013 Behavior and effect of Ti_2Ni phase during processing of NiTi shape memory alloy wire from cast ingot *J Alloy Compd* **581** 344-351
[5] Adharapurapu R R and Vecchio K S 2007 Superelasticity in a new bioimplant material: Ni-rich 55NiTi Alloy *Experimental Mechanics* **47** 365-371
[6] Jiang S, Zhang Y, Zhao Y, Liu S, Hu L and Zhao Ch 2015 Influence of Ni_4Ti_3 precipitates on phase transformation of NiTi shape memory alloy *Trans Nonf Metals Soc of China* **25** 4063-4071

IOP Conf. Series: Materials Science and Engineering **266** (2017) 012009 doi:10.1088/1757-899X/266/1/012009

Microstructure and mechanical properties of selective laser melted Ti6Al4V alloy

M Losertová and V Kubeš

VŠB – Technical University of Ostrava, Faculty of Metallurgy and Materials Engineering, 17. listopadu 15, 708 33 Ostrava-Poruba, Czech Republic

mlosertova@vsb.cz

Abstract. The present work was focused on the properties of porous Ti6Al4V specimens processed by selective laser melting (SLM) and tested in tension and compression before and after heat treatment. The SLM samples were annealed at 955 °C, water quenched and aged at 600 °C with following air cooling. The values of the mechanical tests showed that the samples exhibited high mechanical properties. The anisotropy of tensile and compressive strength was observed, which was related to the occurrence of voids. The plastic properties of specimens were improved by means of the heat treatment that led to the transformation of martensitic to lamellar structure composed of $\alpha + \beta$ phases. The microstructure of SLM samples were evaluated before and after the heat treatment. The brittle nature of failures of non-heat treated samples can be explained by synergy of martensite presence, microcracks and residual stresses produced by SLM.

Keywords: Selective laser melting, Ti6Al4V, Microstructure, Porosity, Mechanical properties.

1. Introduction

The process of the Selective Laser Melting (SLM) represents advantageous production techniques that allows reduction of production steps, a high material use efficiency and a near net shape production [1, 2]. The SLM assures the layer-wise building of parts with a high geometrical complexity and thus, offers broad applications in different industries. However, the final products can displayed some problems related with large thermal gradients occurring in the SLM process due to short interaction times and highly localized input of heat. The influence of the different process parameters as well as of the heat treatments were studied for Ti6Al4V [1-3] that is still the material of particular interest for biocompatible implants in orthopedics and traumatology for its higher biological and biomechanical compatibility in comparison with 316L steel or CoCr alloys [4]. Unlike the parts produced of the compact Ti6Al4V alloy having the Young's modulus equal to 110 GPa [5], the material containing certain level of porosity decreases its modulus of elasticity toward that one of human bones and in surgical implants allows to avoid the stress shielding effect due to the mismatch between the mechanical properties of titanium alloy and bone.

The study of this work was focused on the properties of Ti6Al4V specimens processed by SLM technology and containing certain porosity. The material was tested in tension and compression in the condition with and without heat treatment.

Content from this work may be used under the terms of the Creative Commons Attribution 3.0 licence. Any further distribution of this work must maintain attribution to the author(s) and the title of the work, journal citation and DOI.
Published under licence by IOP Publishing Ltd

2. Experimental

Two sets of samples were built vertically (in a direction *z*), it means in the direction of intended tensile and compressive loading (Figure1), by SLM method in a layer-by-layer manner. The square patterns produced by building strategy of the SLM method are evident in Figure 1a. First set included ten cylindrical tensile test specimens with a gauge length of 30 mm and gauge diameter of 5 mm; second set represented ten cylinders of 10 mm in diameter and 12 mm in height for the compressive tests.

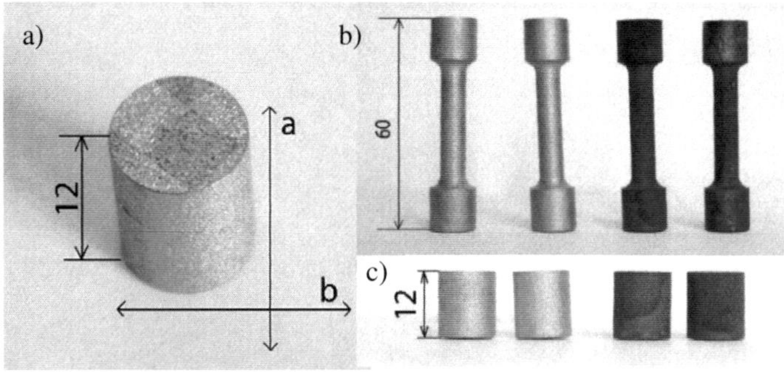

Figure 1. Ti6Al4V specimens built in the direction of height: a) general view with square patterns resulting from building strategy; b) tensile specimens; c) compressive specimens.

Half of the specimens of both sets were processed by heat treatment (HT) under argon atmosphere that consisted of solution annealing at 955 °C followed by furnace cooling to 855 °C and water quenching and ageing at 600 °C followed by air cooling. The surface of heat treated specimens changed the colour to blue-black (Figure 1b and 1c) due to oxidation reactions in argon atmosphere even though the gas purity reached 5N. The thermal regime was chosen in accordance with findings in [3, 6] and is presented in Figure 2.

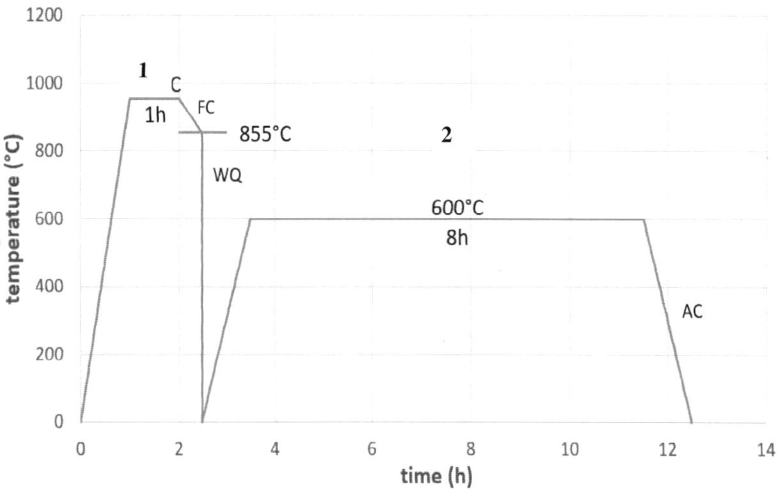

Figure 2. Thermal treatment consisting of two steps: 1) solution annealing followed by furnace cooling (FC) and water quenching (WQ) and 2) ageing followed by air cooling (AC).

The tensile and compression tests at room temperature were performed with strain rate 0.0025 s^{-1} on the specimens with and without heat treatment using Inova Fu-0-350-1700-V1 machine. The porosity was evaluated using image analysis on an optical microscope Olympus IX70. The microstructure features of both sets in different conditions of heat treatment were performed by means of optical microscopy on both cross and longitudinal electrolytic polished sections before and after etching in Kroll's solution (6 % HF, 8 % HNO_3, 86 % H_2O).

3. Results and discussion

The porosity that was determined from ten measurements on five specimens is shown in Figure 3 and reached average values of 5 %. The etched microstructure of non-heat treated specimen is formed of a fine acicular martensite (α′ phase) resulting of the imposed high cooling rate in the SLM process, as seen in Figure 4. The evident building patterns (so called tracks) were processed due to the molten pool boundaries (MPBs) with a special shape in the internal regions and on the surfaces of SLM specimens [7].

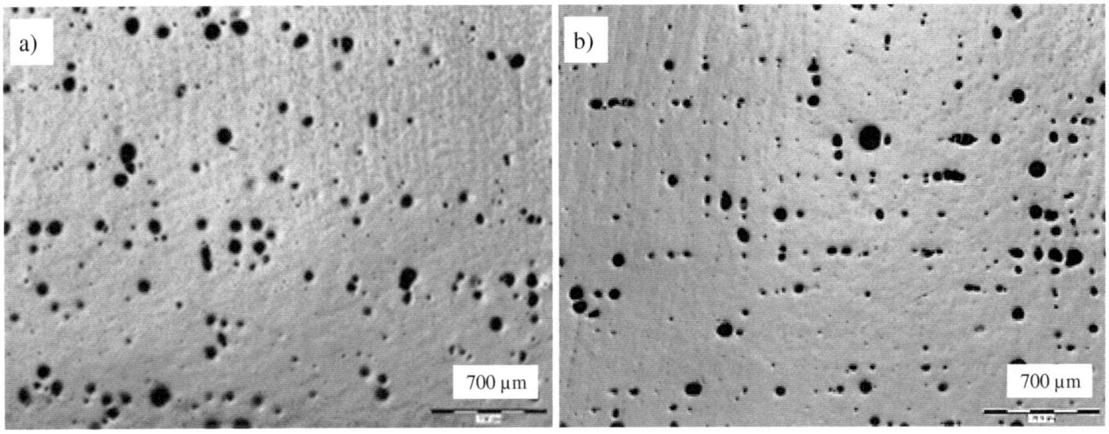

Figure 3. Distribution of the voids in the specimens prepared by SLM and before heat treatment: a) longitudinal and b) cross sections (non-etched).

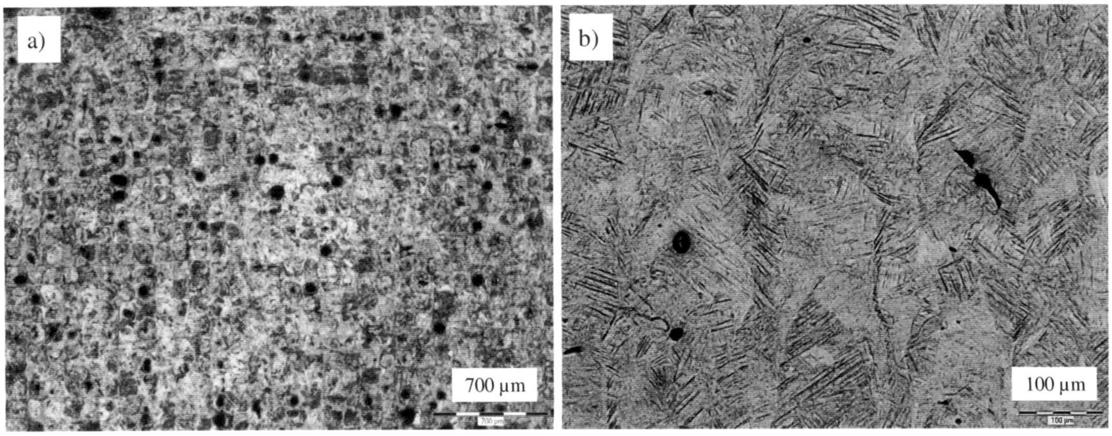

Figure 4. Microstructure of as-built specimen: a) low magnification morphology of building tracks and pores; b) acicular α′ martensite in prior β grains (etched).

After the heat treatment the acicular α′ martensite transformed into equilibrium (α + β) microstructure, at the same time with reducing thermal stresses that have been built up during the SLM process. It is evident that annealing in the vicinity of the β transus (the temperature of the α ↔ β transition is called β transus and for Ti6Al4V reaches 980 °C) led to growing β grains from columnar to large and semi-equiaxed shape that transformed to lamellar (α + β) after the cooling, as seen in Figure 5.

As summarized in Table 1, the mechanical properties in tensile test reached a high yield stress (with average value of 1 056 MPa), a high ultimate tensile strength (1 351 MPa) but a relatively low ductility (up to 6 %). The thermal treatment resulted in slightly increasing yield strength (YS) (by 4 %) but in markedly decreasing ultimate tensile stress (UTS) and elongation (deformation) by 13 and 9 %, respectively. Concerning Young's modulus, the measured values dropped to the 42 and 49 GPa for tensile tested specimens in non-heat treated and heat treated stages, respectively.

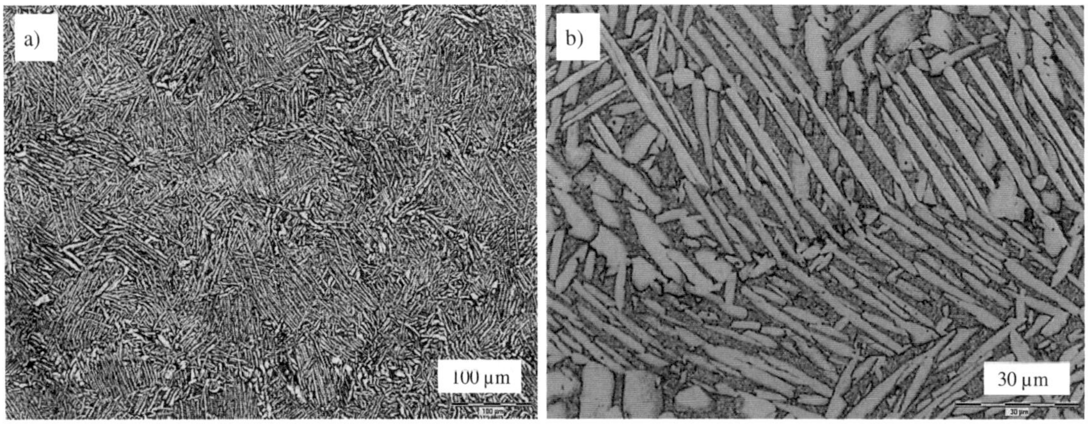

Figure 5. Microstructure of heat treated specimen: a) low magnification morphology of lamellar ($\alpha + \beta$) grains; b) detail of lamellar ($\alpha + \beta$), α is white, β is dark (etched).

Table 1. Mechanical properties of both sets of Ti6Al4V specimens in tensile and compressive testing.

Specimen tested in	Porosity (%)	YS (MPa)	UTS (MPa)	Deformation (%)	E (GPa)
Tensile					
1	5	1060	1312	6	43
2		1083	1373	5.9	44
3		1025	1369	4.5	40
Average value		1056±29	1351±34	5.5±0.8	42±2
6 + HT	5	1055	1184	4.9	49
7 + HT		1102	1171	5.1	50
8 + HT		1137	1189	5	49
Average value		1098±41	1181±9	5±0.1	49±0.6
Compression					
1	5	1099	1763	40	
2		1101	1620	42	
3		1300	1661	23	
Average value		1167±115	1681±74	35±10	
6 + HT	5	1140	1622	25	
7 + HT		1100	1657	27	
8 + HT		1100	1602	27	
Average value		1113±23	1627±28	26±1	

The compression properties achieved of higher values, as seen in Table 1. The heat treatment resulted in decreasing all values of the mechanical properties, by 5, 3 and 16 % for YS, UTS and deformation in compression, respectively. Comparing the results in tensile and compression tests, it is clear that strongly textured structures led to significant anisotropic mechanical behavior. As it can be seen from the Figure 6, the compression properties are higher by 10, 24 and 536 % in values of YS, UTS and deformation, respectively, for as-built specimens, and by 1, 37 and 420 % in YS, UTS and deformation, respectively, for heat treated specimens. The failure surfaces of non-heat treated samples were characterized by brittle nature that can be explained by synergy of martensitic microstructure, microcracks and residual stresses produced by SLM. The lamellar feature of microstructure after heat treatment led to brittle fracture but at lower strength values.

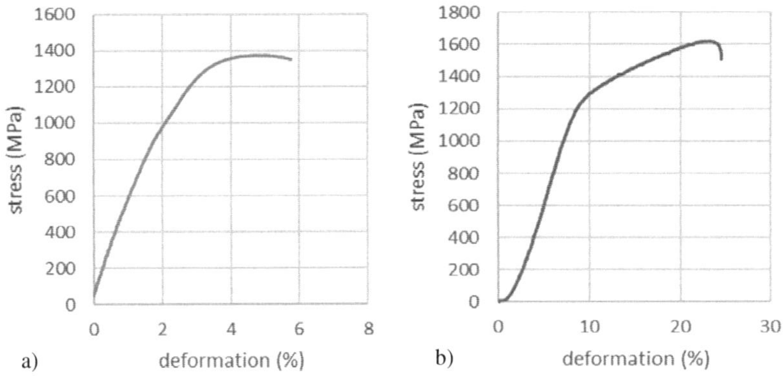

Figure 6. Behavior of the heat treated SLM specimens under loading in a) tension and b) compression.

4. Conclusion

The Ti6Al4V specimens produced by SLM process were mechanical tested in heat treated and non-heat treated conditions. Based on the results of tensile and compression loading the following conclusions can be drawn:

1. Ti6Al4V specimens in the as-built stage consisted of prior-β grains filled with acicular α′ martensite, and usually display high yield strength, but limited ductility.
2. The combined effects of acicular martensite, microcracks and residual stresses were responsible for lower ductility comparing tensile and compressive tests.
3. The martensitic microstructure was the main reason for the higher strength of the as-built specimens.

The high mechanical properties together with lowered Young's modulus are suitable improvement of material properties; however material evaluation of the biocompatible porous alloy is necessary to accomplish by fatigue and corrosion tests in near future.

Acknowledgement

This article has been elaborated in the framework of the project No. LO1203 "Regional Materials Science and Technology Centre - Feasibility Program" funded by Ministry of Education, Youth and Sports of the Czech Republic and co-financed by the European Social Fund and of the projects SGS SP2017/77 and SGS SP2017/58 supported by Ministry of Education, Youth and Sports of the Czech Republic.

References

[1] Sun J, Yang Y and Wang D 2013 Parametric optimization of selective laser melting for forming Ti6Al4V samples by Taguchi method *Optics & Laser Technology* **49** 118-124

IOP Conf. Series: Materials Science and Engineering **266** (2017) 012009 doi:10.1088/1757-899X/266/1/012009

[2] Xu W, Brandt M, Sun S, Elambasseril J, Liu Q, Latham K, Xia K and Qian M 2015 Additive manufacturing of strong and ductile Ti–6Al–4V by selective laser melting via in situ martensite decomposition *Acta Materialia* **85** 74-84
[3] Wauthle R, Vrancken B, Beynaerts B, Jorissen K, Schrooten J, Kruth J-P and Van Humbeeck J 2015 Effects of build orientation and heat treatment on the microstructure and mechanical properties of selective laser melted Ti6Al4V lattice structures *Additive Manufacturing* **5** 77-84
[4] Niinomi M 2008 Mechanical biocompatibilities of titanium alloys for biomedical applications *J Mech Behav Biomed Mater* **1** 30-42
[5] Banerjee R, Nag S and Fraser H L 2005 A novel combinatorial approach to the development of beta titanium alloys for orthopaedic implants *Mater Sci Eng C* **25** 282- 289
[6] Vrancken B, Thijs L, Kruth J-P and Van Humbeeck J 2012 Heat treatment of Ti6Al4V produced by Selective Laser Melting: Microstructure and mechanical properties *J Alloy Compd* **541** 177-185
[7] Shifeng W, Shuai L, Qingsong W, Yan Ch, Sheng Z and Yusheng S 2014 Effect of molten pool boundaries on the mechanical properties of selective laser melting parts *J Mater Process Tech* **214** 2660-2667

IOP Conf. Series: Materials Science and Engineering **266** (2017) 012010 doi:10.1088/1757-899X/266/1/012010

Corrosive effect of environmental change on selected properties of polymer composites

L Markovičová and V Zatkalíková

University of Žilina, Faculty of Mechanical Engineering, Department of Material Engineering, Univerzitná 8215/1, 01026 Žilina, Slovakia

lenka.markovicova@fstroj.uniza.sk

Abstract. The development of composite materials and the related design and manufacturing technologies is one of the most important advances in the history of materials. Composites are multifunctional materials having unprecedented mechanical and physical properties that can be tailored to meet the requirements of a particular application. Ageing is also important and it is defined as the process of deterioration of engineering materials resulting from the combined effects of atmospheric radiation, heat, oxygen, water, micro-organisms and other atmospheric factors. The present article deals with monitoring the changes in the mechanical properties of composites with polymer matrix. The composite was formed from the PA matrix and glass fibers (GF). The composite contains 10, 20 and 30 % of glass fibers. The mechanical properties were evaluated on samples of the composite before and after UV radiation on the sample. Light microscopy was evaluated distribution of glass fibers in the polymer matrix and the presence of cracks caused by UV radiation.

1. Introduction

Modern structural composites, frequently reffered to as advanced composites, can be distinguished from commodity composites because of their frequent use of more exotic or expensive matrix materials and higher-priced reinforcements such as carbon/graphite, and they can be found in more structurally demanding locations that have a greater need for weight savings [1 - 3]. They are a blend of two or more components. One component is made up of stiff, long or short fibers, and the other, for polymeric composites, is a thermoplastic or a resinous binder or matrix that holds the fibers in place.

The fibers are strong and stiff relative to the matrix and are generally orthotropic (having different properties in two different directions) [4 - 7]. These properties are most evident when the components are shown in Figure 1. Polymers are widely used in many industrial fields and in outdoor applications all polymers degrade. The degradation rate depends on the environment (especially sunlight intensity, temperature and humidity) and on the type of polymer. There are some problems such a photo-yellowing and the decrease in mechanical strength after long term irradiation by ultra-violet (UV).

This so-called UV-degradation is due to combined effects of photolysis and oxidative reactions [8]. There are many studies on the photo-degradation of polymers induced by UV irradiation. It is important to study the depth profile of the degradation in order to see the attained depth of the degradation and the effects of the additives.

Content from this work may be used under the terms of the Creative Commons Attribution 3.0 licence. Any further distribution of this work must maintain attribution to the author(s) and the title of the work, journal citation and DOI.
Published under licence by IOP Publishing Ltd

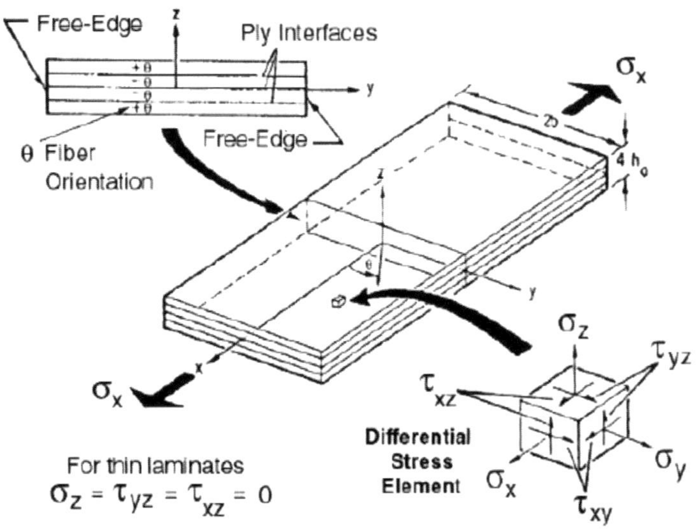

Figure 1. The anatomy of a composites laminate.

Rheology and its experiments reveal information about the flow behaviour of liquids but also the deformation behavior of solids, because it is the typical behaviour of polymers. Changes induced by the environment with degradation effect can be evaluated by rheological measurements which monitor changes in viscoelastic properties of the tested polymers. The fundamental of rheological characteristics is viscosity which defines the internal resistance of material against its creep generated by external forces. It is necessary to realize the dual character of majority of polymer materials from viscoelastic point of view. The action of external force on the ideal viscous material results in its deformation i.e. irreversible locomotion (movement) of macromolecules and after removal of the external force material retains its „new" shape. The action of external force on the ideal elastic material results in its deformation but after the removal of the external force, the material returns to its original shape. Polymers are generally characterized by the viscoelastic nature, which means that external forces cause partly permanent (viscosity element of polymer) and partly reversible (elastic element of polymer) deformation [9, 10].

2. Experimental material

The experimental materials are composites with the PA+PAI (polyamide + polyamidimide) matrix with 10 %, 20 %, 30 % (% = vol. %) of glass fibers coded GF 672 with diameter 10 μm and length 4 mm. The composites are estimated for interior and exterior details of cars, then it is expected their UV stability. The matrix includes also UV stabilizer. The structures of the tested composites were observed by SEM (Scanning Electron Microscopy). The surface of the samples was covered with a carbon layer. In Figure 2, the surface state of the composites with various content of GF is shown and distribution of GF in cross section also. Inhomogenity of the tested composites increases with amount of GF. We can see the fibres are distributed periodically in composite with 10 % of GF (Figure 2a). Higher fibre content (30% GF) causes heterogeneity of the matrix-fibre system (Figure 2c). The individual glass fibres after removal from the matrix, leaving regular oval holes (Figure. 2a, b), with a higher volume of glass fibres is made up of holes.

IOP Conf. Series: Materials Science and Engineering **266** (2017) 012010 doi:10.1088/1757-899X/266/1/012010

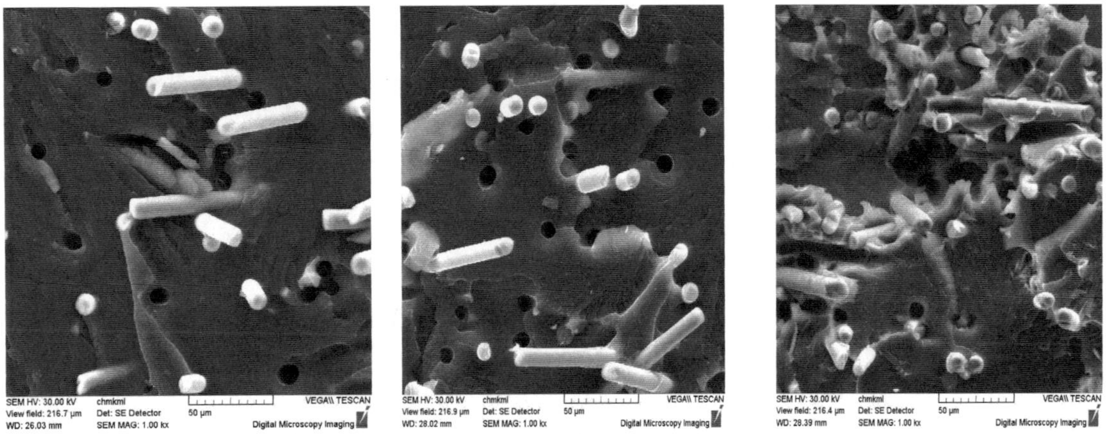

a) 10 % GF b) 20 % GF c) 30 % GF

Figure 2. Structure of composite with glass fibers (SEM).

Testing degradation of polymeric materials is one of the most important tests to the lifetime of polymer product. Ageing tests can be either in real conditions of use of the polymer in a particular application, or using artificial accelerated ageing conditions. Accelerated ageing methods provide test results significantly faster natural aging tests. Testing is based on exposing test bars to man-made climate. After a fixed interval of exposure changes are detected in end points (aesthetic, physical, electrical, etc.). The apparatus for man-made weather ageing ensure continued maintaining of artificial climatic conditions (day and night cycles, changing humidity, drought and wet, etc.). The apparatus SolarBox 1500e which was used in the experiment has the following characteristics: a source of light radiation guarantees a radiant flux of radiation intensity 550 $W.m^{-2}$. The source of light is a xenon arc lamp, but other sources of radiation are allowed too. The device must be equipped with a thermometer built into the black panel, which senses the temperature of the black panel. The black panel temperature of exposure time was selected at 65 °C the liquid phase lasted for 102 minutes and the wet phase for 18 minutes. If it necessary wetting by distilled or deionised water can be applied. The numbers of man-made climate factors that simultaneously affect the test bars is selected by the test program. Test runs continued for a period fixed in the testing program. The duration of the test was 500 and 1000 hours [11].

3. Experimental results and discussion

The Vicat softening temperature is the temperature at which a flat-ended needle penetrates the specimen to the depth of 1 mm under a specific load. The temperature reflects the point of softening to be expected when a material is used in an elevated temperature application. The test specimen must be between 3 and 6.5 mm thick and at least 10 mm in width and length. Of each sample were made three measurements and are determined by the final arithmetic value (Table 1). From the test results, we can see that the Vicat softening temperature slightly increased after 1000 hours of UV radiation.

Table 1. Vicat temperature.

	Vicat temperature [°C]		
Filler content	**0 hrs. UV**	**500 hrs. UV**	**1 000 hrs. UV**
10 %	231	234	243
20 %	240	241	248
30 %	242	243	249

The impact strength test measures the amount of force needed to break a specimen introduced through a swinging pendulum. The samples had the following parameters: width: 10 mm, thickness: 4 mm and a length of 80 mm. The test took place at a temperature of 23 °C (23 ± 2 °C). Results of measurements are shown in Figure 3.

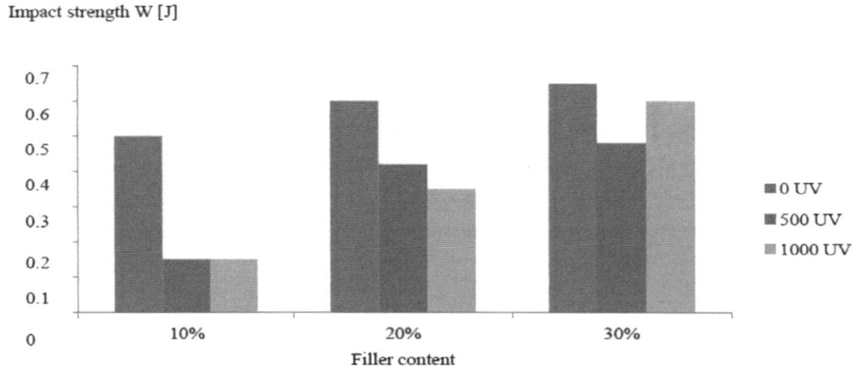

Figure 3. Effect of UV radiation on the change of impact strength.

The results show that the impact resistance to the most impact volume of glass fibers, wherein the amount of the impact strength values increase. The values of impact resistance by the action of UV radiation halved. One of the important properties of the polymer material is the impact strength, which is a combination of strength, plasticity, and durability of the material. When polymers mechanical resistance is dependent on the speed with which a force is applied to the material. At lower speeds the polymer occurs to a relatively large deformation, while at higher speeds has a brittleness of the material, because the mechanical energy is not enough to dispose the material. Because of this feature, the polymer is gradually becoming a better design choice for a variety of applications [12].
Experimental data were obtained by using an oscillatory rheometer Physica MCR301 with convection heating device CTD 450. The main advantage of this measuring method is in simultaneous monitoring of all mentioned parameters, e.g. G′, G″, tg δ and η* [13, 14, 15]:

G´´(Pa) - loss modulus is a measure of deformation energy used up by sample during shear process and shows irreversible deformation energy, presents the viscous behaviour of tested material.

G´(Pa) - storage modulus is a measure of the deformation energy stored by samples during the shear process is showing reversible deformation behaviour, represent elastic behaviour of tested material. The complex modulus is expressed by the equation:

$$|\boldsymbol{G}^*| = \sqrt{(\boldsymbol{G}')^2 + (\boldsymbol{G}'')^2}. \quad (1)$$

η* - complex viscosity represents ratio of η´ expressed elastic behaviour of polymeric and η´´ viscous one [16,17]:

$$|\eta^*| = \sqrt{(\eta')^2 + (\eta'')^2}. \quad (2)$$

The measurements were provided by Frequency Sweep test (FS) to study changes of viscoelastic properties, changes of molecular mass and its distribution after various types and time of degradation. The sample was placed between two parallel plates with diameter of 25 mm, in 1 mm distance from each other. Measurements were carried out at the temperature of 320 °C and following conditions: amplitude of γ = 5%, angular frequency of ω = 500 – 0.051/s. The principle of the mentioned evaluation is shown in Figure 4 [10].

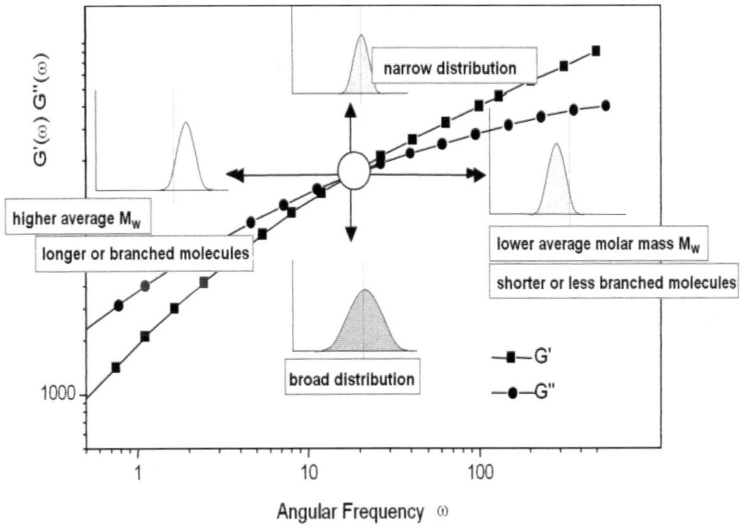

Figure 4. Determination of selected characteristics of material from the curves obtained by FS test [10].

The process of degradation is followed by changes of molecular weight (as networking or macromolecular chains breaking) which should be reflected in the measured parameters. As seen in Figure 4, the values of complex dynamic viscosity η and storage modulus G ' and loss modulus G "can be compared by measuring. Furthermore, we are able to monitor the position of the intersection of the curves characterizing the state of both modulus and indicate the transition from viscous deformation behaviour to the more elastic behaviour. This intersection is qualitative characteristics of the material [16]. The polymer degradation can results in changes of the values of measured characteristics. Shift of the modulus intersection in horizontal direction provides information about the average molecular weight and shift of the modulus intersection in the vertical direction signals the difference in the molecular weight distribution.

Rate and depth of PA + PAI composite with different content of glass fibres degradation after exposure in UV chamber (after 1000 hours) was evaluated by Frequency Sweep test (FS). The differences in viscosity and modulus defining plastic and elastics properties of PA + PAI samples are shown in Figure 5, 6. In Figure 5 it is illustrated difference of rheological parameters of PA + PAI composite with different content of glass fibres (0 hrs. UV radiation) expressed in the position of a cross point (COP) of loss and storage modulus. The COP is moved to lower angular frequencies with increasing molar mass. The vertical shift of the COP indicated changes in distribution of molar mass.

The effect of exposure of the composite PA + PAI simulated atmospheric conditions of intense radiation was clearly observed as a change in viscosity values and as a change of the position of point COP. The viscosity of samples after exposure in the UV chamber shifted towards lower values (Figure 6). It also was recorded by shifting the COP. The intersection of the loss and storage modulus of the samples after exposure to the UV chamber, shifted towards longer and more branched-chain. Differences between visco-elastic properties of specimens in initial state are higher than after exposition. It means increasing of elastic properties and drop of plastic ones. The composites are more brittle, include more cracks and the so the fibres are liberated from their positions.

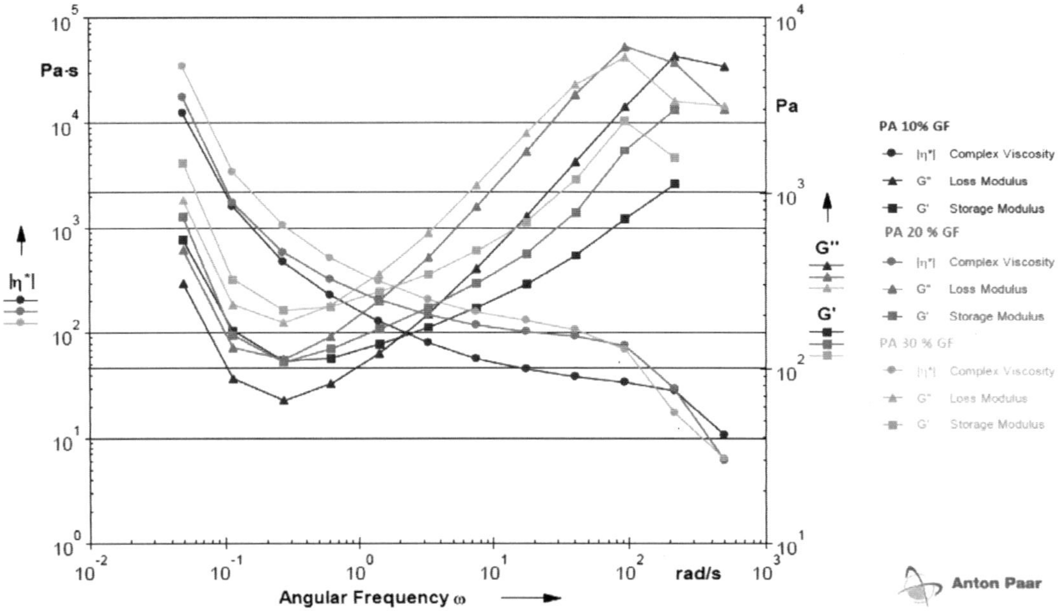

Figure 5. Differences in rheological properties of PA + PAI with different content of GF without UV radiation.

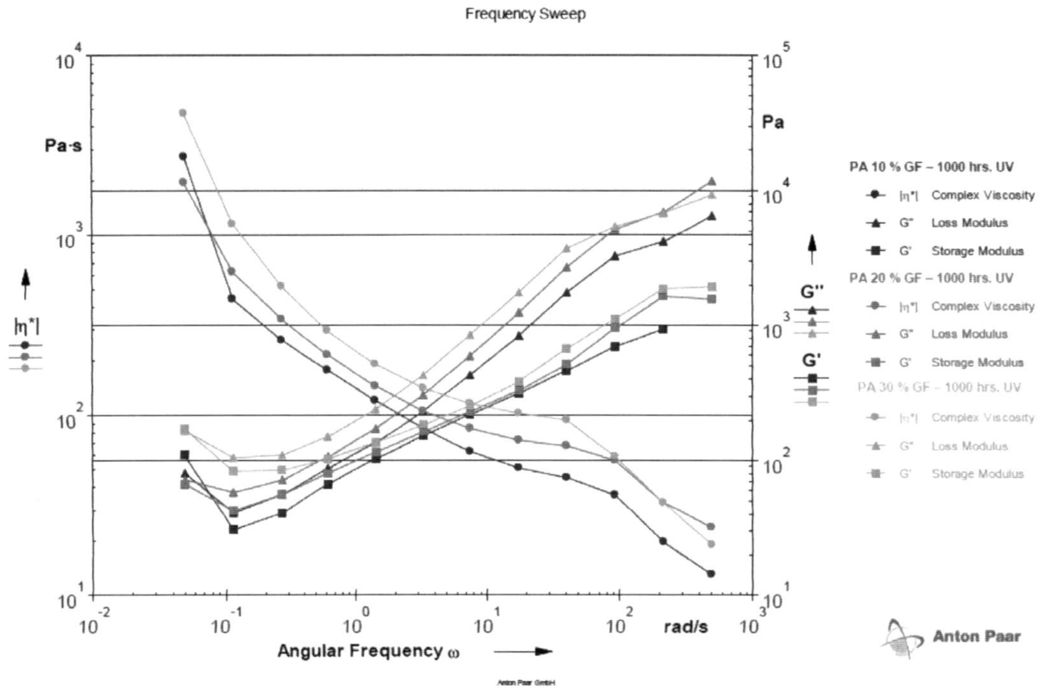

Figure 6. Differences in rheological properties of PA + PAI with different content of GF after 1000 hrs. UV radiation.

4. Conclusions

Based on the experiments performed on composite PA + PAI with different content of the filler (GF) we can conclude:

- From the observation of the morphology of the composite show that the increased volume of glass fibers cause the fibers form clusters and their distribution is uneven. Fracture and irregularities are formed on the surface of the laminate of the matrix observed. The combination of UV radiation and higher glass fiber content rise to cracks, which gradually spread matrix. This caused a degradation of the matrix and the peeling in the form of flakes.
- The softening point of the composite increases with the content of glass fibers. UV radiation has had no significant impact on softening point, there was only a slight change in the material properties. The biggest change in the softening point was the composite containing 10% filler wherein the temperature is increased from 231 °C to 243 °C. This may be caused by changes in the structure and also the partial crosslinking chains.
- In examining the impact strength material with an increasing volume of glass fibers increases. After UV irradiation, this value clearly decreased, although the figures are volatile and vary. This is due to the partial cross-linked polymer by the action of UV radiation and also heterogenous distribution of glass fibers.
- Rheological measurements sensitively monitor changes in the structure of studied polymers due to degradation effect in different environment. This was reflected in the measured rheological parameters and variables like viscosity, molecular weight changes and their distribution, modulus characterizing elastic and plastic properties.
- UV radiation causes - decrease in viscosity values, the formation of long or branched chains, confirming the displacement of the COP point in regions with a broad molecular weight distribution.
- The changes by the effect of environment with present UV radiation reduce products quality and life time. The possibility of reducing degradation processes is in using of effective UV stabilizers and increase adhesiveness of GF to matrix.

Acknowledgement

The research was supported partially by Slovak Academy of Science grant VEGA No. 1/0533/15 and No. 1/0123/15.

References

[1] Peters S T, Humphrey W D and Foral R F 1999 Filament Winding *Composite Structure Fabrication* 2nd ed. SAMPE Publishers Covina.
[2] Brown R T 1987 *Engineered Materials Handbook, vol. 1 Composites* Theodore Reinhart Tech. Chairman ASM International.
[3] Birley A W, Haworth B and Batchelor J 1992, *Physics of Plastic* Carl Hanser Verlag Munich.
[4] Gupta P K 1988 Glass Fibers for Composite Materials *Fibre Reinforcements for Composite Materials* A. R. Bunsell Ed. Elsevier Publishers.
[5] Petru M, Broncek J, Lepsik P and Novak O 2014 Experimental and Numerical Analysis of Crack Propagation in Light Composite Materials Under Dynamic Fracturing *Communications - Scientific Letters of the University of Zilina* **16** (3A) 82-89

[6] Dresslerova Z, Palcek P and Uhricik M 2016 Effect of Heat Treatment on the Behavior of AZ31 and AZ91 Magnesium Alloys During the Cyclic Loading *Proc. of Materials Today* **3** (4) 965-968

[7] Raz K, Zahalka M and Chval Z 2017 Injection molding quality improvement by advanced virtual simulations *Manufacturing Technology* **17** ISSN 1213-2789 79-83

[8] Denisov E T 1982 In: Scott G. Editor Developments in polymer stabilization **5** London Applied Science 23-40

[9] Carraher Ch E Jr 2007 *Introduction to Polymer Chemistry* Taylor and Francis Group ISBN 0-8493-7047-2 USA

[10] Wollny K 2006 Comparison of a propylene reinceforced with glass fiber and pure polypropylen over a temperature range from -150 to +180°C Anton Paar Germany GmbH

[11] Markovicova L. Zatkalikova V and Vasko A 2016 Accelerated Aging of Polymeric Composites in Laboratory Conditions *Manufacturing Technology* **6** (5) 1033-1037

[12] Kanie T, Fujii K, Arikawa H and Inoue K 2000 Flexural properties and impact strength of denture base polymer reinforced with woven glass fibers Dental materials **16** Issue 2 150-158

[13] Mezger T G 2006 The Rheology Handbook 2 Hannover Vincentz Network ISBN 3-87870-174-8 299

[14] Vojsovičová M, Liptáková T and Zatkalíková V 2011 Rheological characteristic of thermoplastic polymer after degradation *In Communications Scientific Letters of the University of Zilina* **13** EDIS Publishing Institution of Zilina University ISSN 1335-4205 32-35

[15] Lelovics H and Liptáková T 2010 Rheological properties of acrylic bone cement SmartSet®HV In *Communications Scientific Letters of the University of Zilina* **12** EDIS Publishing Institution of Zilina University ISSN 1335-4205 85-89

[16] Carraher C E 1996 Polymer Chemistry An Introduction 4/e Marcel Dekker Inc. New York 53

Polyethylene glycol (PEG) used in the preparation of (Er^{3+}/Yb^{3+}):$LiNbO_3$ waveguides

D Mikolášová[1], K Rubešová[1], V Jakeš[1], P Nekvindová[1], L Nádherný[1] and J Oswald[2]

[1] University of Chemistry and Technology, Department of Inorganic Chemistry, Technická 5, 166 28 Prague 6, Czech Republic
[2] Academy of Sciences of the Czech Republic, Institute of Physics, Cukrovarnická 10, 162 00 Prague 6, Czech Republic

mikolasd@vscht.cz

Abstract. $LiNbO_3$ co-doped with Er^{3+} and Yb^{3+} belongs among the most promising materials in photonics. In the form of thin films on a suitable substrate, it can be used as an active waveguide. Only few articles have been published concerning the deposition of such thin layers using a sol-gel method. Our aim was to test polyethylene glycol (PEG) as a water-soluble polymer in solutions to be deposited by spin-coating on sapphire substrates. Two heat treatment regimes were tested after single layers were deposited. Photoluminescence in a range of 1450–1650 nm was measured and the final films exhibited emission spectra with 6 local maxima. The deposited (Er^{3+}/Yb^{3+}):$LiNbO_3$ films were also characterized by m-line spectroscopy. In the NIR range, the films behaved as one-mode waveguides. This non-hydrolytic sol-gel method based on PEG thus represents one of few methods applicable in $LiNbO_3$ waveguiding films preparation.

1. Introduction

Due to its extraordinary spectrum of physical properties, $LiNbO_3$ is one of the most widely used dielectric materials [1]. Besides, $LiNbO_3$ is a very important material in optics and photonics – whether pure or doped with d- or f-ions [2]. Co-doping with Er^{3+} and Yb^{3+} ions results in the additional photoluminescence of the $LiNbO_3$ material. The energy of the radiative transition $^4I_{13/2} \rightarrow ^4I_{15/2}$ of erbium f-electrons (1450–1650 nm with a maximum of 1530 nm) is well applicable in the so called third telecommunication window. (Er^{3+}/Yb^{3+}):$LiNbO_3$ in the form of a waveguide can be then employed as an amplifier or switcher. To prepare a waveguiding structure, various techniques are used – either the modification of a $LiNbO_3$ single crystal or the deposition of $LiNbO_3$ thin film. In the single crystals, a waveguide can be created by doping ions to the surface layer – e. g. high-energy ion implantation or high-temperature diffusion [3, 4]. To deposit (Er^{3+}/Yb^{3+}):$LiNbO_3$ thin films, physical (e. g. Pulsed Laser Deposition or Liquid Phase Epitaxy) [5, 6] or chemical methods belonging to the sol-gel family [7] are used.

The sol-gel method is generally a cost-effective procedure that is able to transfer elemental homogeneity from a solution up to a solid sample. Thin films can be deposited by spray drying, spin- or dip-coating using either pre-hydrolysed sols (hydrolytic alkoxide methods) or homogeneous solutions (chelation or polymeric processes). Lithium niobate in the form of thin films has been prepared with several methods [7, 8]; however, high demands on microstructure quality cause that

Content from this work may be used under the terms of the Creative Commons Attribution 3.0 licence. Any further distribution of this work must maintain attribution to the author(s) and the title of the work, journal citation and DOI.
Published under licence by IOP Publishing Ltd

very few articles describe the preparation of thin layers able to guide actually an optical signal [9, 10]. We were successful previously with the use of polyvinylpyrrolidone (PVP) as a chelating and anti-stress agent at the deposition of waveguiding (Er^{3+}/Yb^{3+}):$LiNbO_3$ layers [11, 12]. Microstructure is not the only limitation for waveguiding properties. Films must also be of sufficient thickness and have high refractive index contrast with a substrate for an optical signal guiding. This can be limiting in a sol-gel method compared to the physical methods of films deposition. However, PVP can overcome this problem and layers with a thickness over 1 μm can be prepared [13]. Not so in the case of $LiNbO_3$ – the presence of the highly hydrolyzing Nb^{+V} ion does not allow for the preparation of solutions concentrated sufficiently to produce films of a thickness over 1 μm. In spite of all these factors, our (Er^{3+}/Yb^{3+}):$LiNbO_3$ films were proven as active waveguides and therefore we decided to continue in testing of another water-soluble polymer – polyethylene glycol (PEG).

Based on our previous study, we tested one- and two-step decomposition regime after each layer deposition. After crystallization, the films were characterized with the aim to test its effect on luminescent and waveguiding properties of prepared (Er^{3+}/Yb^{3+}):$LiNbO_3$ structures.

2. Experimental

The preparation of a Li^{+I}/Nb^{+V}/Er^{+III}/Yb^{+III} solution was carried out in an anhydrous and inert Ar atmosphere. First, $Nb(OC_2H_5)_5$ (STREM Chemicals; 99.9 %) was dissolved in 2-methoxyethanol (2-ME; Sigma-Aldrich; anhydrous; 99.8 %) and refluxed at 125 °C for 1 hour. Then, polyethylene glycol (PEG; Sigma-Aldrich; M_w~200 g/mol; purity not specified) was added to the Nb(V) solution and further refluxed at 125 °C for 1 hour. Finally, lithium acetate (Sigma-Aldrich; 99.5 %), erbium acetate (Sigma-Aldrich; 99.9 %) and ytterbium acetate (Sigma-Aldrich; 99.9 %) were added and the solution was refluxed at 125 °C for another 1 hour. Prior to the preparation, the acetates were thermally dehydrated at 200 °C [14-16]. $Er(Ac)_3$ and $Yb(Ac)_3$ were added in a concentration of 0.5 at% each with respect to the $LiNbO_3$ stoichiometry. The final Li^{+I}/Nb^{+V}/Er^{+III}/Yb^{+III} solution had a metal concentration of 0.6 mol/L and the molar ratio of Li:Nb:PEG:2-ME was equal to 1:1:1:15.

The films were prepared by spin-coating on sapphire (0001) substrates (sized 1.5 cm × 1.5 cm). The coater WS-650SZ-6NPP/LITE with the 650 controller unit was utilized for the spin-coating (Laurell Technologies Corporation). The optimized deposition parameters were: rotation at 5000 rpm for 30 s and the volume of the injected solution 55 μl; a total of 5 layers were deposited on each substrate. The heat treatment of each deposited layer was performed in one-step (700 °C/10 min in air) or two-steps regime (350 °C/10 min in air and 700 °C/10 min in air). The preparation and names of thin films samples are summarized in Table 1.

Table 1. The names and heat treatment condition of prepared thin films.

	PEG-1S	**PEG-2S**
PEG molecular mass (g/mol)	200	200
number of deposited layer	5	5
heat treatment in air	one-step (700 °C/10 min)	two-step (350 °C/10 min and 700 °C/10 min)

Thermal behavior of deposited solutions was studied by differential thermal analysis (DTA) and thermogravimetric analysis (TG) using the SetSys Evolution equipment (Setaram). The DTA and TG data were measured in a range of 20–800 °C with a heating rate of 5 °C/min in air. The phase composition of the prepared samples was determined by XRD. The device used was Bruker-AXS D2 Phaser powder diffractometer with the parafocusing Bragg–Brentano geometry using the CoK_α radiation. Data were evaluated in the software package HighScore Plus. The film morphology was characterized by atomic force microscopy (AFM) using Ntegra Spectra (NT-MDT). The surface scans were performed in a tapping (semi-contact) mode. Cantilevers with a strain constant of 1.5 kN m^{-1}

equipped with a standard silicon tip with curvature radius lower than 10 nm was used for all measurements.

The photoluminescence spectra of the prepared thin films were collected within a range of 1440–1600 nm at room temperature. A semiconductor laser POL 4300 emitting at 980 nm was used for the excitation of electrons. The luminescence radiation was detected by a two-step-cooled Ge detector J16 (Teledyne Judson Technologies). To scoop specific wavelengths, a double monochromator SDL-1 (LOMO) was used. A synchronous detection technique was implemented by chopping the laser beam at a modulation frequency of about 35 Hz and by employing a lock-in amplifier (EG&G 5205). For the evaluation, all luminescence spectra were transformed to the base level and normalized to a standard sample – an $Er^{3+}/Yb^{3+}:LiNbO_3$ monocrystalline sample. Transmission spectra were collected in a range of 300–1800 nm using two different equipments. In a range of 300–1100 nm, a fiber spectrometer Avaspec-3648 (Avantes) was used. In a wavelength range of 600–1800 nm, the transmission was measured by a single grating spectrometer MDR 23 with tungsten lamp. The spectra measured by the two spectrometers were linked up at 600 nm. The film thickness was determined by a UV-vis reflectance spectrometer Avaspec-3648 (Avantes). Waveguiding properties were measured by m-line mode spectroscopy using the Metricon Prism Coupler device. Our measurement was done using a prism with an effective range of refractive index from 1.55 to 2.45 (for 633 nm) at operating wavelengths of 633, 964, 1311 and 1552 nm in the TE polarization.

3. Results and discussion

Based on DTA/TG measurement (Figure 1), we suggested the thermal treatment of deposited layers. The first endothermic step between 20 °C and 130 °C belongs to the evaporation of 2-methoxyethanol. The following two exothermic processes (in this case with no significant mass change) can be asigned to the decomposition of present organics – acetates, 2-methoxylesters of acetic acid and PEG. As in the case of PVP, we decided to compare one- and two-step heating at 700 °C and 300 °C/700 °C, respectively.

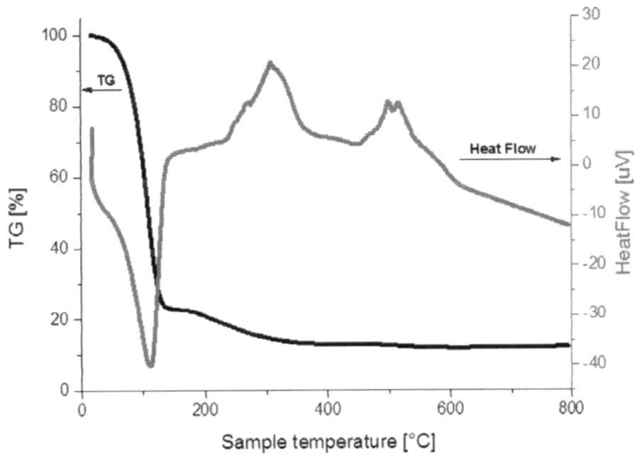

Figure 1. DTA/TG measurement of the $Li^{+I}/Nb^{+V}/Er^{+III}/Yb^{+III}$ precursor solution.

Figure 2 shows X-ray diffraction patterns of the prepared films. The measurement is presented only in a range of 20–47° without the dominating sapphire (0001) reflection at 48.7°. The $LiNbO_3$ reflections were indexed according to the PDF 00-020-0631; the reflections of the sapphire substrate are marked by *. Prepared films were polycrystalline with no secondary phase detectable. The films exhibited the preferential orientation of $LiNbO_3$ along the (006) plane; however, also other reflections were present – (012), (104), (110) and (113). In comparison with our previous sol-gel prepared $LiNbO_3$ films [11, 12], the described disorientation is more significant; however, this phenomenon does not have any major influence on the waveguiding ability of thin films (as described below).

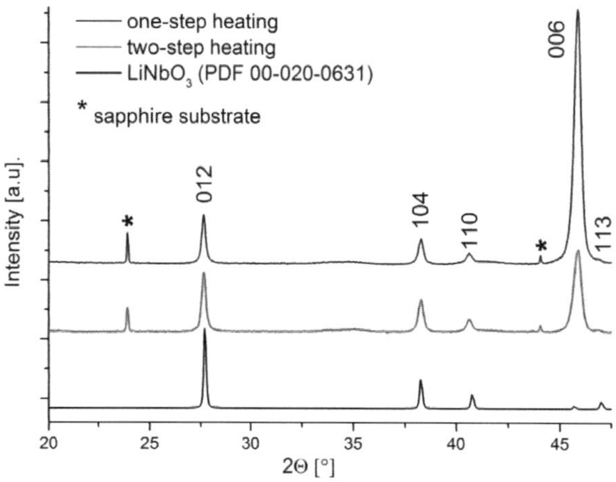

Figure 2. XRD patterns of thin films prepared using different heat treatment of deposited layers. A simulated pattern of $LiNbO_3$ using PDF 00-020-0631 is also presented. (for a black white version: in the legend, the order from top to bottom is the same as the order of lines in the picture)

The photoluminescence spectra were measured in a wavelength range of 1440–1650 nm where the $^4I_{13/2} \rightarrow {}^4I_{15/2}$ transition of Er^{3+} ions is located. Figure 3 presents the emission spectra of both samples (the intensity was normalized to a standard $Er^{3+}/Yb^{3+}:LiNbO_3$ monocrystalline sample). The heat treatment used did not have any influence on the structure of the spectra, only a slight difference in the intensity is visible. In both cases, the most intense band is placed at 1530 nm, other distinguishable bands are placed at 1485 nm, 1510 nm, 1543 nm, 1558 nm (and one broad band around 1600 nm). Such energy separation of emission spectrum is indicative of a proper crystal surrounding of Er^{3+} ions and of overall good crystallinity of the $LiNbO_3$ phase. In comparison with the sole other published $Er^{3+}/Yb^{3+}:LiNbO_3$ thin film prepared using a sol-gel method [17] (besides our articles using PVP), the clear bands position in the emission spectra is comparable to monocrystalline thin films [3].

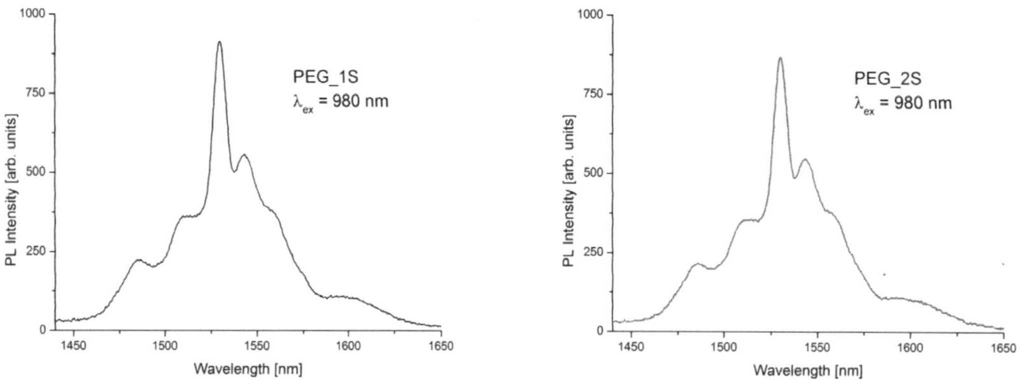

Figure 3. Photoluminescence emission spectra excited using a 980 nm laser measured at room temperature. The intensity was normalized to a standard $Er^{3+}/Yb^{3+}:LiNbO_3$ monocrystalline sample.

Regarding the waveguiding properties of prepared thin films, m-line spectroscopy was performed at the following wavelengths: 633 nm, 964 nm, 1311 nm and 1552 nm. Table 2 summarizes the number of modes guided by both samples.

Table 2. The summary of optical modes guided, and the thickness and roughness of prepared films.

Sample	Wave propagation (number of modes)				Thickness	R_a
	633 nm	964 nm	1311 nm	1552 nm	[nm]	[nm]
PEG-1S	0	1	1	1	771	3.29
PEG-2S	0	1	1	1	743	3.98

In accordance with our previous experience with polycrystalline $LiNbO_3$ thin films, the waveguiding ability disappears at shorter wavelengths where an optical signal is probably dispersed and attenuated by grains with a size comparable to the wavelengths used. This assumption can be supported by transmittance spectra in the VIS and NIR region (Figure 4) where the interaction of a film with the light is increased when wavelength decreases. The roughness and thickness of the films was comparable - Table 2 (taking into account the precision of the roughness determination using AFM images). The AFM images of both samples are presented in Figure 5. Contrary to monocrystalline samples, the waveguiding ability in polycrystalline $LiNbO_3$ samples is strongly driven by microstructure; thickness is a minor, but not legible, parameter [18]. In any case, one-mode $Er^{3+}/Yb^{3+}:LiNbO_3$ active waveguides were prepared.

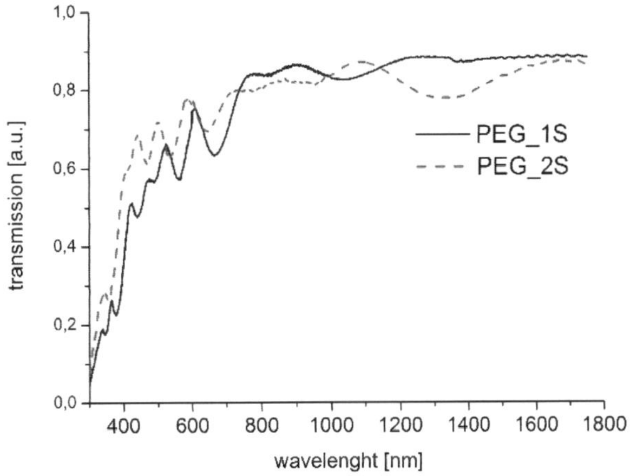

Figure 4. Transmission spectra of both prepared films measured in a spectral range of 300–1800 nm.

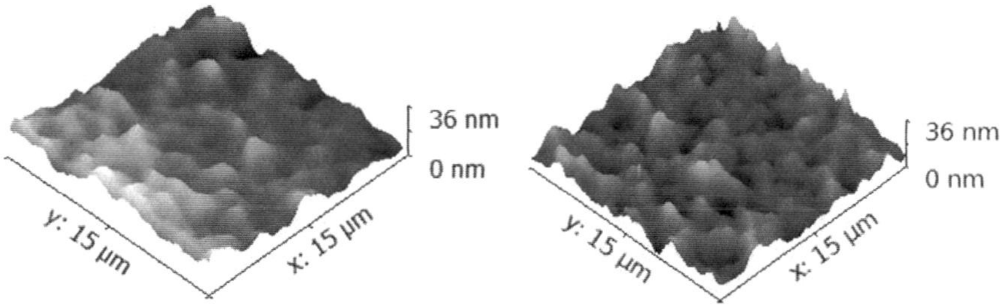

PEG-1S PEG-2S

Figure 5. AFM images of prepared thin films. The calculated roughness is presented in Table 2.

4. Conclusions

We present Er^{3+}/Yb^{3+}:$LiNbO_3$ waveguiding thin films deposited by the spin-coating of solutions based on polyethylene glycol. Two different heat treatments of deposited layers were applied and the influence on the optical properties was established. Irrespective to the treatment used, one-mode waveguides were prepared. These results, together with our previous films prepared with another water-soluble polymer – PVP, are promising for the development of active waveguide technology based on $LiNbO_3$ material. Especially the high homogeneity of dopant ions, demonstrated here by photoluminescence spectra, is crucial in testing new dopants. Comparing this sol-gel method with physical deposition processes, the thickness of deposited films, achievable by sol-gel methods, can with difficulty reach 1 μm without loss of suitable microstructure. On the other hand, this method represents a cost-effective and simple method applicable in $LiNbO_3$ thin films deposition.

Acknowledgements

This work was financially supported by specific university research (MSMT No 20-SVV/2017) and by the GA CR, project No. P108/12/G108.

References

[1] Wong K K 2002 *Properties of lithium niobate* (INSPEC, IEE)
[2] Arizmendi L 2004 *Physica Status Solidi (a)* **201** 253-83
[3] Cajzl J., Nekvindová P, Macková A, Malinský P, Oswald J, Staněk S, Vytykáčová S and Špirková J 2016 *Optical Materials* **53** 160-8
[4] Nekvindova P, Svecova B, Cajzl J, Mackova A, Malinsky P, Oswald J, Kolistsch A and Spirkova J 2012 *Optical Materials* **34** 652-9
[5] Jelínek M, Oswald J, Kocourek T, Rubešová K, Nekvindová P, Chvostová D, Dejneka A, Železny V, Studnička V and Jurek K 2013 *Laser Physics* **23** 105819
[6] Kaigawa K, Kawaguchi T, Imaeda M, Sakai H and Fukuda T 1997 *Journal of Crystal Growth* **177** 217-25
[7] Hirano S., Takeichi Y, Sakamoto W and Yogo T 2002 *Journal of Crystal Growth* **237–239, Part 3** 2091-7
[8] Bouquet V., Leite E R, Longo E, Varela J A, Viry M G and Perrin A 2001 *Journal of the European Ceramic Society* **21** 1521-4
[9] Nashimoto K, Moriyama H and Osakabe E 1996 *Japanese Journal of Applied Physics Part 1-Regular Papers Short Notes & Review Papers* **35** 4936-40
[10] Ono S. and Hirano S I 2007 *Journal of the Ceramic Society of Japan* **115** 801-7
[11] Rubešová K, Mikolášová D, Hlásek T, Jakeš V, Nekvindová P, Bouša D and Oswald J 2016 *Journal of Luminescence* **176** 260-5
[12] Rubešová K, Mikolášová D, Hlásek T, Jakeš V, Nekvindová P, Matějka P, Dendisová M, Zlámalová Cílová Z and Oswald J 2017 *Journal of Physics and Chemistry of Solids* **111** 343
[13] Kozuka H and Kajimura M 2000 *Journal of the American Ceramic Society* **83** 1056-62
[14] Casado F J M, Riesco M R, Redondo M I, Choquesillo-Lazarte D, Lopez-Andres S and Cheda J A R 2011 *Crystal Growth & Design* **11** 1021-32
[15] Hussein G A M 2001 *Powder Technology* **118** 285-90
[16] Hussein G A M and Balboul B A A 1999 *Powder Technology* **103** 156-64
[17] Takahashi M, Yoshiga T, Kajitani N, Takeda Y, Sato S, Wakita K, Ohnishi N, Hotta K and Kurachi M 2006 *Japanese Journal of Applied Physics* **45** 7305-10
[18] Hewig G M , Jain K, Sequeda F O, Tom R and Wang P-W 1982 *Thin Solid Films* **88** 67-74

IOP Conf. Series: Materials Science and Engineering **266** (2017) 012012 doi:10.1088/1757-899X/266/1/012012

A novel approach to quantitative analysis of the local deformation in grained structure

S Minárik[1] **and M Martinkovič**[2]

[1]Slovak University of Technology in Bratislava, Faculty of Materials Science and Technology in Trnava, Advanced Technologies Research Institute, Ulica Jána Bottu č. 2781/25, 917 24 Trnava, Slovakia
[2]Slovak University of Technology in Bratislava, Faculty of Materials Science and Technology in Trnava, Institute of Production Technologies, 917 24 Trnava, Slovakia

stanislav.minarik@stuba.sk

Abstract. Novel technique for identifying and quantification of local deformation phenomena in continuum model of grained solid structure is presented in this paper. We propose a method that combines deformation properties of structure with the changes of grains geometry deformed under the influence of local force. Experimental analyses of grains shapes subjected to a local force show a specific spatially oriented prolongation of grains in direction determined by applied force. However the volume of each grain is retained during the force application in case of plastic deformation. Character of individual grain prolongation depends on the initial shape of grain and direction of loading force. There is a definite relationship between the change in grain shape and nature of the driving deformation force. We contribute to the revealing of mentioned relationship because of we propose and analyze a method for quantification of the effect of grain shape modification on the basis of grain deformation. Quantitative analysis of local deformation in grained structure can be realized in a perspective using mentioned method. Map of local deformation data in grained system can be constructed in this way and next the local deformation dynamics analysis can be performed. However precision of mentioned analysis must be proven by evaluating of its practical predictive performance in future.

1. Introduction

Various technological procedures in the materials preparation are influenced by macroscopic effects of deformation. However these observable macroscopic effects do not reflect local microstructural changes in deformed areas of material in all their diversity. To obtain exact information about local changes in microstructure various more sophisticated methods must be used, for instance methods based on acoustic emission, microhardness measure, dislocations analysis, slip band observation, micromeshes method or macroscopic screw method. But no of these available methods give analytical expressions relating structure parameters and local strain in each position of bulk material volume.

The polycrystalline material is formed from grains having a various shapes. During an applied mechanical loading, according to the different directions and the influences from neighbouring grains, most of grains are submitted to a deformation. The deformation of grain results in a change in orientation and amount of grain surface area which lead to local heterogeneous plasticity. Therefore

Content from this work may be used under the terms of the Creative Commons Attribution 3.0 licence. Any further distribution of this work must maintain attribution to the author(s) and the title of the work, journal citation and DOI.
Published under licence by IOP Publishing Ltd

IOP Conf. Series: Materials Science and Engineering **266** (2017) 012012 doi:10.1088/1757-899X/266/1/012012

grain surface area and grain edge per unit volume are parameters important in various theories of plasticity as they are generally considered to heterogeneous nucleation sites.

Our work is oriented just on the mathematical treatment of grained structure deformation. Contribution deals with the problem of a grained structure distortion induced by local deformation. We suggest a variable suitable for evaluation of degree of grain surface area orientation with respect to certain vector in three-dimensional space and discuss the relation of this variable to local deformation parameters. Our contribution finishes with a brief commentary on suggested technique of quantitative analysis.

We emphasize that strictly geometric orientation of the grain in three-dimensional space is discussed in our work. We do not discuss orientation of the crystal structure within the grain (even though concept of grain orientation is usually related with crystallographic orientation). A new magnitude κ is defined to evaluate the degree of orientation of the grain surface with respect to a given direction. The presented work aims to find a correlation between the change in the degree of geometrical orientation of a grain and the deformation experienced by it. We study the case of a grain subjected to deformation, and apply the obtained results to the particular case of a randomly oriented grained structure. It is necessary to note that presented paper contains only a mathematical model for the describing the deformation of a fictional, simplified grained structure - no real material is specified, nor is the actual deformation process described. Purely theoretical approach is presented. However it have been shown that such a deformation mechanism, i.e. elongation (or the shortening) of grains under applied load, exists in a real-world material.

2. Quantitative analysis of grains orientation

Grain-surfaces (grain boundaries) play important role in energy balance of grained structure. Loading force causes a specific type of perturbation of this balance in initial grained structure. Therefore it may be supposed that in case of grain deformation certain direction is preferred from the orientation of grain-surface elements point of view. Question is how could be quantified relationship between this preferred orientation of grain-surface area elements and local structure deformation parameters.

Planar character of infinitesimal surface area enables to clearly define orientation of surface element in 3D just by means of the vector $d\vec{S}$. Hence it is possible to easily analyse a sort of "degree" of single surface element orientation in relation to any unit vector $\vec{\tau}$ for example by absolute value of scalar product:

$$\left|\vec{\tau}\cdot d\vec{S}\right| = dS\left|cos(\alpha)\right| = dS_p \tag{2.1}$$

where α is angle between vectors $\vec{\tau}$ and $d\vec{S}$. This absolute value determines the size of projection of the surface element dS_p to the plane perpendicular to the vector $\vec{\tau}$. Quantity (2.1) calculated per unit surface area of grain, i.e.:

$$\frac{dS\left|cos(\alpha)\right|}{S}, \tag{2.2}$$

assesses a contribution of the element to a "degree" of whole grain-surface orientation in relation to vector $\vec{\tau}$. We suggest, that for quantitative analysis of grain surface distortion it is required to define a variable κ enabling evaluate "degree" of whole grain-surface orientation in relation to vector $\vec{\tau}$.

Contribution rate of mentioned element to the whole grain-surface orientation can be intuitively proposed by generalization of formula (2.2) as:

$$d\kappa = \frac{dS}{S}\xi(\alpha),$$

where S is total surface of whole grain and $\xi(\alpha)$ is a function of the angle α. It is advisable to choose the function $\xi(a)$ in such a way that the value κ will be related to an experimentally measurable quantities. We have chosen function $\xi(\alpha)$ in such a way that limit values of variable κ are consistent

IOP Conf. Series: Materials Science and Engineering **266** (2017) 012012 doi:10.1088/1757-899X/266/1/012012

with values obtained in procedures realized at analysis of particles orientation using Saltykov method [1, 2]. Experimental estimation of grains orientation degree using this method is relatively simple. In consistent with this technique we require:

$$\lim_{\alpha\to\pi/2} \kappa = 1 \ , \qquad \lim_{\alpha\to 0} \kappa = -1 \ . \tag{2.3}$$

So next function can be intuitively used in this case:

$$\zeta(\alpha) = 1 - 2\left|\cos(\alpha)\right| \tag{2.4}$$

and contribution $d\kappa$ can be written in the form:

$$d\kappa = \left\{1 - 2\left|\cos(\alpha)\right|\right\}\frac{dS}{S} = \frac{dS}{S} - 2\frac{\left|\vec{\tau}\cdot d\vec{S}\right|}{S} \ . \tag{2.5}$$

Degree of whole grain-surface orientation in relation to vector $\vec{\tau}$ can be determined by the integrating of (2.5) over the entire grains surface:

$$\kappa = 1 - 2\frac{1}{S}\int_{(Surface)} \left|\vec{\tau}\cdot d\vec{S}\right| \ . \tag{2.6}$$

We believe that variable κ defined by (2.6) enables to analyze grain-surface area distortion induced by deformation.

3. Relationship between deformation tensor and orientation parameter κ

Consider isotropic, linear elastic material with the grained structure. Generally the parametrization of any grain boundary surface in 3D can be accomplished by spatial coordinates of each points on the surface:

$$X_i = X_i(u,v), \text{ where } i = 1, 2, 3. \tag{3.1}$$

and u, v are parameters that changes in defined intervals. Therefore the position vectors of each points on the surface of single grain boundary with any possible shape in undeformed configurations can be written as:

$$\vec{R}(u,v) = \left[X_1(u,v), X_2(u,v), X_3(u,v)\right] \ , \quad u \in \langle u_1, u_2\rangle \ , \quad v \in \langle v_1, v_2\rangle . \tag{3.2}$$

Let the position vectors of each points on the surface of the same grain in deformed configurations are:

$$\vec{r}(u,v) = \left[x_1(u,v), x_2(u,v), x_3(u,v)\right] \ , \quad u \in \langle u_1', u_2'\rangle \ , \quad v \in \langle v_1', v_2'\rangle . \tag{3.3}$$

In general case, any 3D deformation can be described by a 3x3 deformation tensor $\bar{\bar{\varepsilon}}$. Using notation mentioned above the grain linear deformation (for small strain) can be expressed as anisomorphic linear transformation:

$$\vec{r} = \bar{\bar{\varepsilon}}\cdot\vec{R}, \quad \text{or} \quad x_i = \varepsilon_{ij} X_j, \tag{3.4}$$

where i, j = 1, 2, 3 and ε_{ij} are coefficients of deformation tensor. It is no problem to verify that $\varepsilon_{ij} = \delta_{ij}$ in undeformed case (where δ_{ij} is Kronecker symbol). We used typical Einstein's summation convention in the second formula in (3.4) whereby when an index variable appears twice in a single term it implies summation of that term over all the values of the index.

Considering transformation (3.4) when calculating the orientation parameter (2.6) we get:

$$\kappa_{defor} = 1 - 2\frac{\int\limits_{(Surface)} \left| \left(\vec{\vec{\sigma}}^T \cdot \vec{\tau} \right) \cdot d\vec{S}_{undefor} \right|}{\int\limits_{(Surface)} \left| \vec{\vec{\sigma}} \cdot \vec{\rho} \right| dS_{undefor}} \tag{3.5}$$

where $\vec{\vec{\sigma}}$ is a tensor coefficients of which are related to coefficients of deformation tensor $\vec{\vec{\varepsilon}}$:

$$\sigma_{wm} = (-1)^{w+m} \det\left(\vec{\vec{\varepsilon}}_{mw}\right) . \tag{3.6}$$

$\det\left(\vec{\vec{\varepsilon}}_{mw}\right)$ is the minor (or subdeterminant) of matrix of the deformation tensor $\vec{\vec{\varepsilon}}$. So it is easily to show that coefficients of tensor of grain surface deformation can be found as:

$$\sigma_{wm} = \varepsilon_{ik}\varepsilon_{jn} - \varepsilon_{in}\varepsilon_{jk} , \tag{3.7}$$

We note that $\vec{\vec{\sigma}}^T$ in formula (3.5) is tensor that is transposed to the tensor $\vec{\vec{\sigma}}$ determined by (3.6). It can be seen from (3.5) that in general the transformation of parameter κ during grain deformation appears to be relatively complicated and we will not examine its mathematical properties in detail. However, it should be noted that we integrate over the surface of undeformed grain in the formula (3.5). So to determine the value κ in deformed configuration we need to know only the shape of undeformed grain and coefficients of deformation tensor ε_{ij}. Therefore we believe, that κ could be an acceptable scalar variable suitable for the quantitative evaluation of grain-surface distortion during local deformation of grained structure.

4. Monte Carlo simulation of the parameter κ - the case of "random" grained structure

Integrals in formula (3.5) must be integrated over the entire surface of grains boundaries if the whole grained structure orientation κ is evaluated. For the calculation purposes this integral can be transformed to summation over the small areas placed around whole grains boundaries surface (infinitely small in the ideal case). So we get:

$$\kappa_{defor} \approx 1 - 2\frac{\sum_{i=1}^{N} \left| \left(\vec{\vec{\sigma}}^T \cdot \vec{\tau} \right) \cdot \vec{s}^{(i)} \right|}{\sum_{i=1}^{N} \left| \vec{\vec{\sigma}} \cdot \vec{\rho}^{(i)} \right| s^{(i)}} . \tag{4.1}$$

where $\vec{\tau} = [\tau_1, \tau_2, \tau_3]$ is any unit vector (i.e. $\tau_1^2 + \tau_2^2 + \tau_3^2 = 1$) and $\vec{\rho}^{(i)}$ is unit vector perpendicular to the i-th grain-surface element of undeformed grain:

$$\vec{s}^{(i)} = \vec{\rho}^{(i)} s^{(i)} . \tag{4.2}$$

Vector $\vec{\rho}^{(i)}$ can be written in the form:

$$\vec{\rho}^{(i)} = \left[\sin\left(\eta^{(i)}\right)\cos\left(\chi^{(i)}\right), \sin\left(\eta^{(i)}\right)\sin\left(\chi^{(i)}\right), \cos\left(\eta^{(i)}\right) \right] , \quad \chi^{(i)} \in \langle 0, 2\pi \rangle, \quad \eta^{(i)} \in \langle 0, \pi \rangle$$

while angles $\eta^{(i)}$ and $\chi^{(i)}$ determines orientation of the i-th grain-surface element and then:

$$\vec{s}^{(i)} = \left[s_1^{(i)}, s_2^{(i)}, s_3^{(i)} \right] = \left[s^{(i)} \sin\left(\eta^{(i)}\right)\cos\left(\chi^{(i)}\right), s^{(i)} \sin\left(\eta^{(i)}\right)\sin\left(\chi^{(i)}\right), s^{(i)} \cos\left(\eta^{(i)}\right) \right] . \tag{4.3}$$

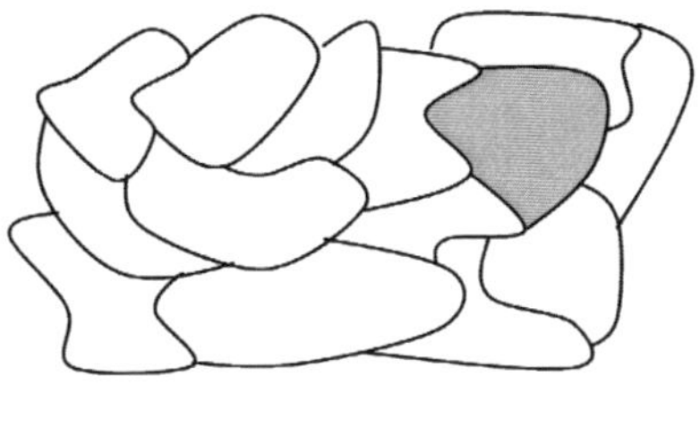

Figure 1. Scheme illustrating a view on a grained structure cross-section (for example on metallographic cut). Cross-section of some individual grain is colored.

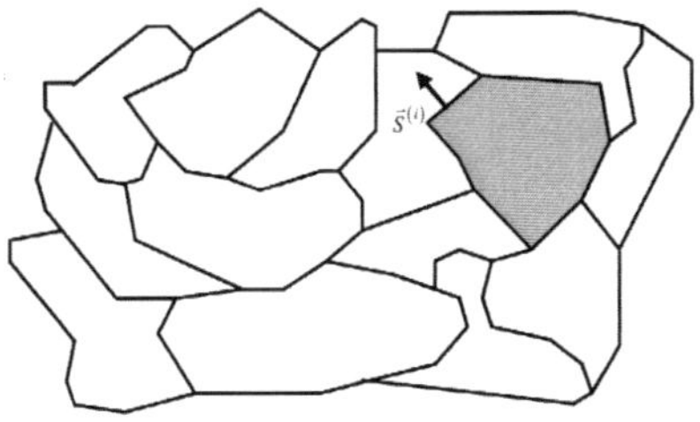

Figure 2. Scheme illustrating a view on the model of the same grained structure cross-section. Scheme demonstrates the replacement of grain surface by small flat areas.

If we use result (3.7) and notations (4.3) sums in the formula (4.1) can be easy calculated. As can be shown the problem of grained structure orientation can be formulated mathematically as random matrix problems. In this sense, the formula (4.1) can be rewritten to the form:

$$\kappa_{defor} \approx 1 - 2\frac{\sum_{i=1}^{N}\left(S^{(i)} \left| \sum_{n=1}^{3} \tau_n \det \tilde{A}_n^{(i)} \right| \right)}{\sum_{i=1}^{N}\left(S^{(i)} \sqrt{\sum_{n=1}^{3}\left(\det \tilde{A}_n^{(i)} \right)^2} \right)} \qquad \ldots \quad n = 1, 2, 3 \tag{4.4}$$

where $\det \tilde{A}_n^{(i)}$ is determinant of random matrix $\tilde{A}_n^{(i)}$ with coefficients:

$$\tilde{A}_{nmk}^{(i)} = \varepsilon_{mk}\left(1-\delta_{m,n}\right) + \frac{1}{2}\delta_{m,n}\Big\{\left[1+(-1)^{\alpha_k}\right]\sin\left(\eta^{(i)}\right)\cos\left(\chi^{(i)}\right) + \left[1+(-1)^{k}\right]\sin\left(\eta^{(i)}\right)\sin\left(\chi^{(i)}\right) + \left[1+(-1)^{\beta_k}\right]\cos\left(\eta^{(i)}\right)\Big\} \tag{4.5}$$

and:

$$\alpha_k = \sum_{\nu=1}^{k}\frac{1+(-1)^{\nu}}{2}, \quad \beta_k = \sum_{\nu=1}^{k}\frac{1-(-1)^{\nu}}{2}, \quad \ldots \quad n,m,k = \{1, 2, 3\}.$$

Formula (4.4) can be imposed and smartly used in the formal mathematical notation as well as numerical implementation of considered approximation approach.

We suppose calculate degree of whole grained structure orientation κ in relation to any unit vector $\vec{\tau}$ by means of the formula (4.4) while sums in the numerator and denominator in the second member on the right side of this formula can be determined step by step. Any randomly oriented surface element of grain boundary $S^{(i)}$ in the given bulk sample with grained structure can be included into the calculation of κ if corresponding parameters $\eta^{(i)}$ and $\chi^{(i)}$ will be applied. Therefore random values of angles $\eta^{(i)}$ and $\chi^{(i)}$ should be substituted step by step into formula (4.5) when the calculation progress.

5. Investigation of the simulation procedure

In this section we present investigation of computational stability of procedure described above. Consider random grained structure with uniform distribution of parameters $\eta^{(i)}$ and $\chi^{(i)}$. Values of these parameters were generated by random number generator character of which was tested (see Figure 3). As can be seen from the Figure 3 no direction is preferred from the flat element $S^{(i)}$ orientation point of view. Let the scalar variable $\kappa = 0$ for this grained structure in undeformed state. Consider for example the deformation of mentioned structure described by next deformation tensor:

$$\vec{\vec{\varepsilon}}_{defor} = \begin{pmatrix} 0.8\,, & 1.2\,, & 1.45 \\ 0.42\,, & 0.5\,, & 0.62 \\ 1.09\,, & 2.22\,, & 0.722 \end{pmatrix}. \qquad (5.1)$$

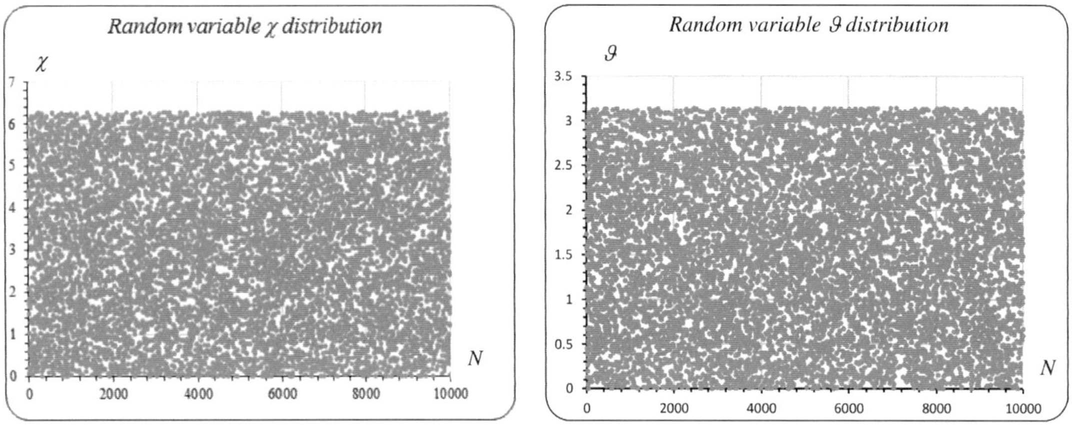

Figure 3. Random number generator test. Each point on graphs corresponds to generated value of the parameter.

Degrees of whole grained structure orientation κ_{defor} in relation to a few arbitrary chosen vectors $\vec{\tau}$ were calculated by Monte Carlo simulation using the formula (4.4).
Results of the simulation are shown in the Figure 4. As can be seen from presented graphs if the number of considered flat elements N is increasing the variance of calculating values κ_{defor} decreases. Computational test was carried out up to $N = 10^8$.

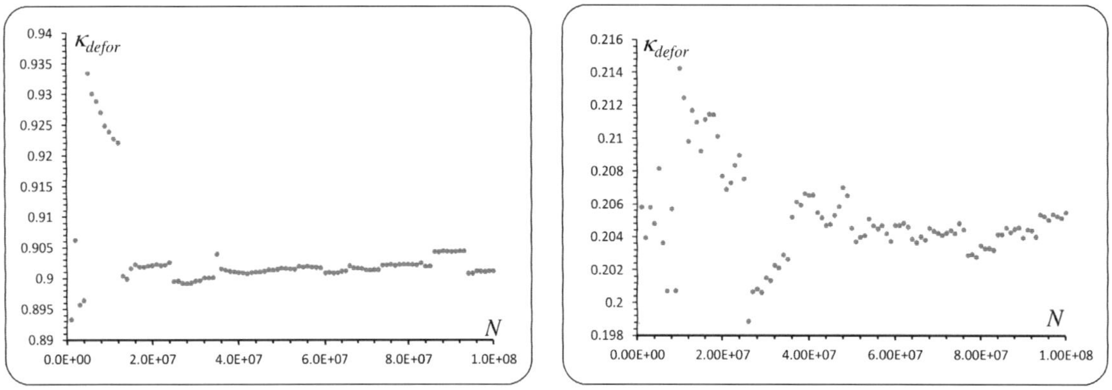

a) $\vec{\tau} = [0,0,1]$, limit value $\kappa_{defor} = 0.9013$ b) $\vec{\tau} = [1,0,0]$, limit value $\kappa_{defor} = 0.2055$

IOP Conf. Series: Materials Science and Engineering **266** (2017) 012012 doi:10.1088/1757-899X/266/1/012012

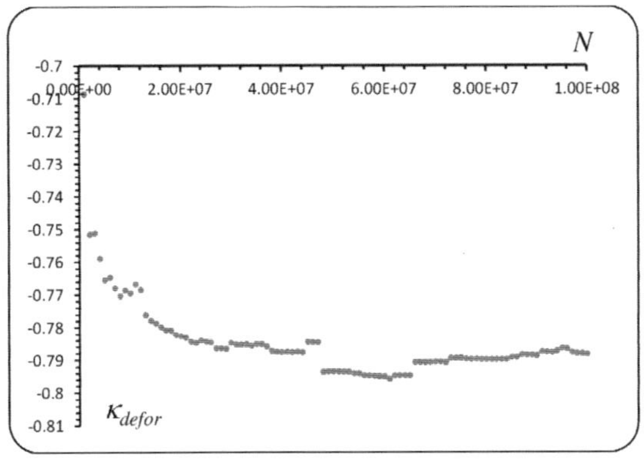

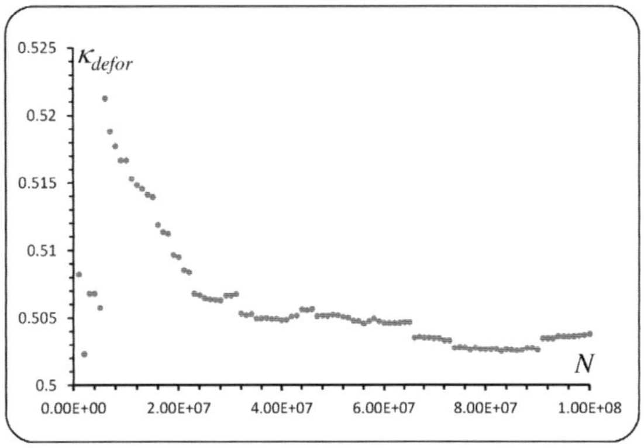

c) $\vec{\tau}=[0,1,0]$, limit value $\kappa_{defor} = -0.78816$ d) $\vec{\tau}=[0.5,0.5,0.70711]$, limit value $\kappa_{defor} = -0.5038$

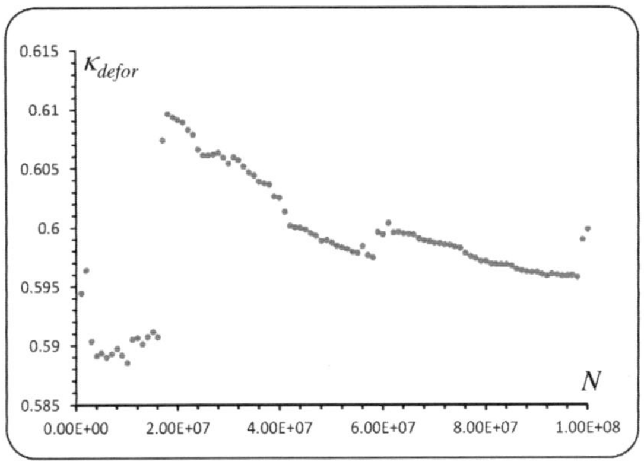

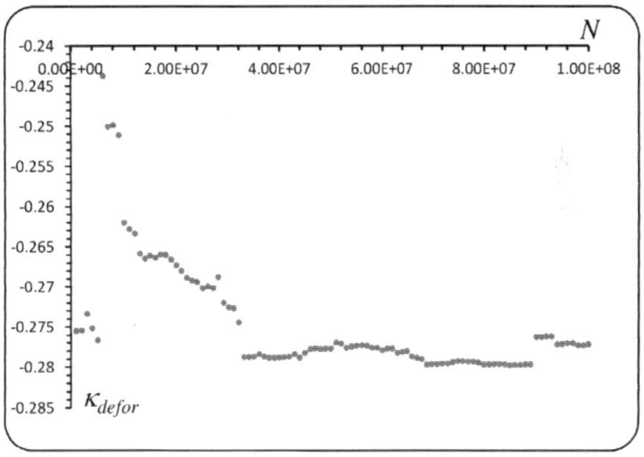

e) $\vec{\tau}=[0.171,0.296,0.94]$, limit value $\kappa_{defor} = 0.599$ f) $\vec{\tau}=[0,0.707,0.707]$, limit value $\kappa_{defor} = -0.277$

Figure 4. Result obtained by Monte Carlo simulation of parameter κ for the deformed grained structure described by deformation tensor (5.1). Graphs show calculated value of parameters κ_{defor} as a function of considered number of randomly oriented grains boundaries elements N in the stochastic model of grained structure illustrated in the Figure 2. Presented results clearly demonstrate the convergence of calculated values κ_{defor} when the simulation is running. This fact suggests the applicability of proposed procedure for investigation of local deformation in real grained structures.

6. Conclusion

In this paper values of parameteters established in stereology are used to extend the results of stereological measurements to provide a predicts the local deformation of grained structure. Orientation of grains in a microstructure reflects this deformation in terms of quantities that can be directly measured in microstructures by means of stereological methods. This is illustrated with a set of simulations that successfully tests our predictions.

Generally one can easily understand that there is a correlation between the change in the grain orientation and grain deformation [3, 4]. This correlation was mathematically demonstrated in our work. Indeed, the expression (3.5) allows one to investigate this correlation. Finally, if we interpret the quantity κ as the degree of grain orientation corresponding with the value measured at Saltykov method, the analysis of local deformation could be easily realized experimentally on the basis of our results. As an illustration we have shown the applicability of our results for random model of grained structure. This specific case of model have used here because of its simplicity but we believe that our method can be applied also for other models.

Acknowledgment
This work was supported by the VEGA Grant No. 1/0122/16 of the Grant Agency of the Slovak Republic Ministry of Education and by the Slovak Research and Development Agency under the contract No. APVV-15-0319.
This work was supported by the project No. 003STU-2-3/2016 named "Development of the Research Infrastructure in SUT" funded from government budget of Slovak Republic by the grant from the Ministry of Education, Science, Research and Sport of the Slovak Republic.

References

[1] Saltykov S A 1970 Stereometric metallography *Metallurgia* Moscow (in Russian)
[2] Lewis H D, Walters K L and Johnson K A 1973 Particle size distribution by area analysis: Modifications and extensions of the Saltykov method *Metallography* 6(2) 93–10
[3] Maier B, Mihailova B, Paulmann C, Ihringer J, Gospodinov M, Stosch R, Güttler B and Bismayer U 2009 Effect of local elastic strain on the structure of Pb-based relaxors: A comparative study of pure and Ba- and Bi-doped PbSc0.5Nb0.5O3 *Physical Review* B(79) 224108
[4] Xu Y H, Pitot H C 2003 An improved stereologic method for three-dimensional estimation of particle size distribution from observations in two dimensions and its application. *Comput Methods Programs Biomed* 72(1) 1-20

IOP Conf. Series: Materials Science and Engineering **266** (2017) 012013 doi:10.1088/1757-899X/266/1/012013

Design of welding parameters for laser welding of thin-walled stainless steel tubes using numerical simulation

M Nagy and M Behúlová

Slovak University of Technology in Bratislava, Faculty of Materials Science and Technology in Trnava, Ulica Jána Bottu č. 2781/25, 917 24 Trnava, Slovakia

mate.nagy@stuba.sk

Abstract. Nowadays, the laser technology is used in a wide spectrum of applications, especially in engineering, electronics, medicine, automotive, aeronautic or military industries. In the field of mechanical engineering, the laser technology reaches the biggest increase in the automotive industry, mainly due to the introduction of automation utilizing 5-axial movements. Modelling and numerical simulation of laser welding processes has been exploited with many advantages for the investigation of physical principles and complex phenomena connected with this joining technology. The paper is focused on the application of numerical simulation to the design of welding parameters for the circumferential laser welding of thin-walled exhaust pipes from theAISI 304 steel for automotive industry. Using the developed and experimentally verified simulation model for laser welding of tubes, the influence of welding parameters including the laser velocity from 30 $mm.s^{-1}$ to 60 $mm.s^{-1}$ and the laser power from 500 W to 1200 W on the temperature fields and dimensions of fusion zone was investigated using the program code ANSYS. Based on obtained results, the welding schedule for the laser beam welding of thin-walled tubes from the AISI 304 steel was suggested.

1. Introduction

The laser beam welding represents a very powerful, productive and efficient technology of joining of various materials, which is used in many industries, from the automotive, aviation, space industries through electronics, medicine to energetic or military industries [1-10]. Modelling and numerical simulation of laser welding processes has been exploited with many advantages for the investigation of physical principles and complex phenomena connected with this joining technology [11-16]. It is generally considered as an effective tool for definition of appropriate welding parameters and optimization of laser welding processes [3, 17-18].

Initial experiments in the field of welding of thin-walled tubes from the AISI 304 stainless steel by solid-state lasers have been carried out in 1974 mainly for the needs of the energy industry in the Europe and America. Up to the present, a large number of reactors and heat exchangers for European nuclear power plants are made from the AISI 304 steel. This steel is also widely used in the automotive industry around the world [5, 19-20].

The main objective of this paper is to design proper welding parameters for the production of weld joints of thin-walled exhaust system tubes from the AISI 304 steel using the solid-state disk laser. To predict the temperature fields in welded pipes, numerical simulation of the laser beam welding in the program code ANSYS is exploited.

Content from this work may be used under the terms of the Creative Commons Attribution 3.0 licence. Any further distribution of this work must maintain attribution to the author(s) and the title of the work, journal citation and DOI.
Published under licence by IOP Publishing Ltd

2. Problem description

The tubes with the inner diameter of 27.1 mm, the thickness of 0.8 mm and the length of 200 mm from the AISI 304 stainless steel with the chemical composition given in the Table 1 should be welded using the solid-state disk laser TRUMPF TruDisk 4002 with the wavelength of 1063 nm, the fiber diameter of 200 μm and the maximum power of 2 kW [21]. Argon gas with the maximum flow rate of 30 l.min^{-1} can be used as a shielding gas to protect both, weld bead and weld root. To design the appropriate laser power for required welding speed from 30 mm.s^{-1} to 60 mm.s^{-1}, numerical simulation of the laser welding was performed using the program code ANSYS [22].

Table 1. Chemical composition of the AISI 304 steel.

Element	C	Cr	Ni	Mn	Si	P	S	Fe
wt. [%]	max. 0.08	18 - 20	8 - 10.5	max. 2.0	max. 1.0	0.045	0.03	bal.

3. Simulation model

In general, numerical modelling is understood as a set of research activities associated with the development of appropriate geometrical model for an investigated object, setting its material properties, definition of relations between the modelled object and surroundings and also verification and validation of created simulation model for the analysed process or system. Hence, the simulation model consists of several parts including geometrical, finite element and material model. An important part constitutes the assignment of loads and boundary conditions. In case or transient processes, the definition of initial conditions is also required.

Numerical simulation of laser beam welding processes requires to solve a coupled thermal, fluid and stress-strain problem involving phase transformations [11, 23-25]. Transient temperature fields developed during the fusion welding can be described by the heat diffusion partial differential equation in the form [26]

$$\rho c_p \frac{\partial T}{\partial t} = \left[\frac{\partial T}{\partial x}\left(\lambda_x \frac{\partial T}{\partial x}\right) + \frac{\partial T}{\partial y}\left(\lambda_y \frac{\partial T}{\partial y}\right) + \frac{\partial T}{\partial z}\left(\lambda_z \frac{\partial T}{\partial z}\right) \right] + q_v. \tag{1}$$

where T is the temperature [°C], t is the time [s], ρ is the density [kg.m^{-3}], c_p is the specific heat [J.kg^{-1}.K^{-1}], λ is the thermal conductivity [W.m^{-1}.K^{-1}] in the x, y, z directions of Cartesian coordinate system and q_v represents the heat generated in the unit volume of material per unit time [W.m^{-3}], i. e. the volumetric density of internal heat sources due to the welding heat input. To solve the Fourier-Kirchhoff partial differential equation (1), it is necessary to specify geometrical shapes and dimensions of welded components, thermo-physical properties of welded materials, initial temperature distribution at the beginning of a welding process, intensity of internal heat sources and the thermal effects at the interface of welded structures and environment.

According to the symmetry of investigated welding process and the main aim of numerical simulation (setting of welding parameters), only a one half of a welded tube with the length of 20 mm was taken into account. The designed geometrical model is illustrated in Figure 1a.

To perform the transient thermal analysis, 3D finite element mesh (Figures 1b and 1c) was generated according to supposed temperature gradients using the eight-node 3D element type of SOLID 70 [22]. That means that progressive meshing with the higher elements density in the region of the weld pool and in the heat affected zone (HAZ) was used. The length of elements along the welding line was 0.2 mm. Through the tube thickness, twelve elements were generated for the accurate modelling of the volumetric heat source (Figure 1d) and thus, attaining reasonable temperature distribution in the region of a weld joint. The overall number of elements and nodes was 66 647 and 65 022, respectively.

IOP Conf. Series: Materials Science and Engineering **266** (2017) 012013 doi:10.1088/1757-899X/266/1/012013

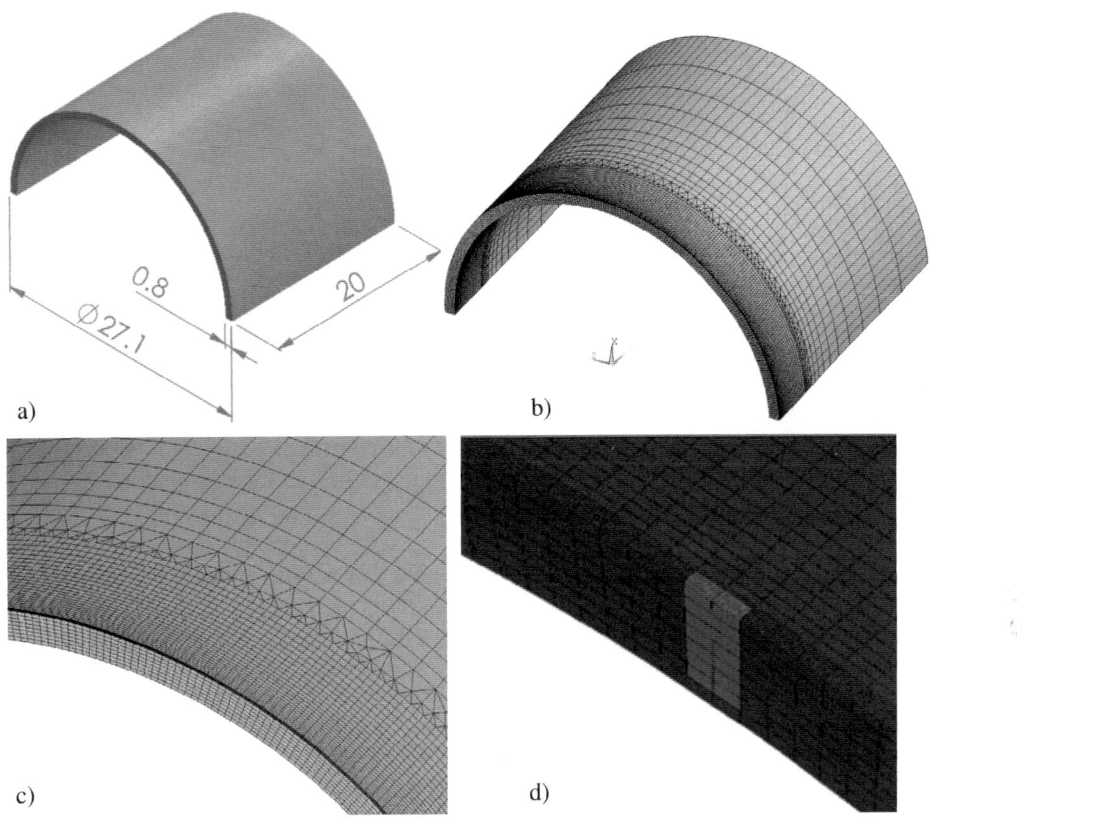

Figure 1. Geometrical model (a) and FE model (b) with a detail of generated mesh (c) and applied volumetric heat source (d).

The thermal properties of the AISI 304 steel in the dependence on temperature (Figure 2) were computed using the JMatPro software [27]. The equilibrium solidus and liquidus temperatures for this steel were computed to be 1360 °C and 1440 °C, respectively. The enthalpy of fusion was modelled applying the method of modified specific heat [28-29].

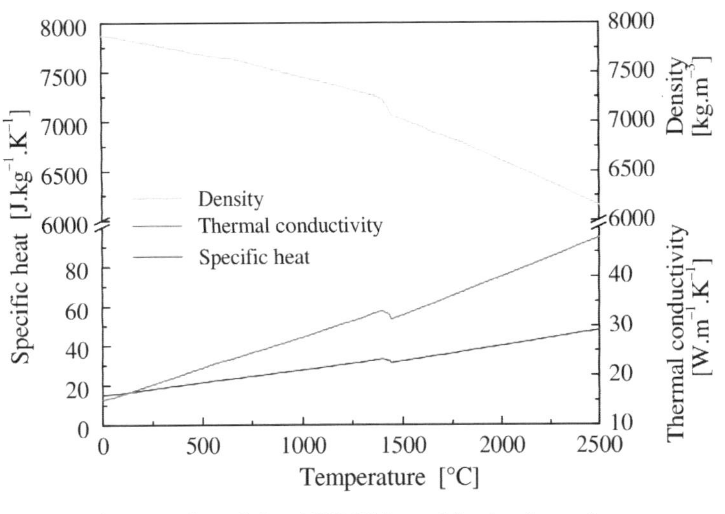

Figure 2. Thermal properties of the AISI 304 steel in the dependence on temperature.

IOP Conf. Series: Materials Science and Engineering **266** (2017) 012013 doi:10.1088/1757-899X/266/1/012013

Considering the small thickness of welded tubes and the low thermal conductivity of AISI 304 steel, a simple volumetric heat source model was applied for simulation of laser beam welding. The volumetric density of internal heat sources q_v [W.m^{-3}] was computed by the relationship

$$q_v = \frac{\eta P}{V} \tag{2}$$

where P is the laser power [W], V represents the volume of a heat source [m^3] and η is the efficiency. Depending on the applied laser (mainly the laser wavelength and power), welded material, its phase state, surface quality, temperature and other parameters, the laser energy irradiating the surface of welded material can be partially absorbed, reflected and transmitted. The process efficiency is mainly affected by absorption of laser energy. Based on the literature data, the absorption coefficient for the AISI 304 stainless steel was established to be 37.4 % [30].

The initial temperature of welded tubes was supposed to be 20 °C. The cooling of welded components by convection and radiation to the ambient air and the argon shielding gas was neglected in this case as the heat extraction from the weld bead to the surrounding tube base material by conduction is significantly more important.

4. Verification of simulation model by experimental laser welding

Experimental laser beam welding of thin-walled tubes from AISI 304 stainless steel with the inner diameter of 25.5 mm, the thickness of 0.8 mm and the length of 200 mm was carried out using a solid state DY044 Nd:YAG diode laser from ROFIN-SINAR Laser GmbH company as the disk laser TRUMPF TruDisk 4002 was not available for experimental measurements. The basic technical parameters of the DY044 Nd:YAG diode laser are summarized in Table 2.

Table 2. Technical parameters of a solid state DY044 Nd:YAG diode laser.

Parameter	Value	Units
Maximum power	4.4	kW
Wave length	1053	μm
Beam quality	12	mm.mrad
Laser beam diameter	400	μm

The welded tubes were clamped into a rotary chuck. The welding speed was limited by the maximal speed of applied dog rotary chulk which was 17 mm.s^{-1}. To investigate the influence of welding speed on the quality of produced weld joints, the welding speeds of 10 mm.s^{-1}, 14 mm.s^{-1} and 17 mm.s^{-1} were used. The laser power was varied from 400 W to 1000 W. During experimental laser beam welding, the temperatures were measured by two thermocouples of the K type. The thermocouples were welded by hand on the both sides of weld joints. In this reason, the location of thermocouples was not the same for all samples. The distance of thermocouples from the weld centerline were a (thermocouple TC1) and b (thermocouple TC2) according to Figure 3a. The welding parameters exploited for the preparation of experimental weld joints are given in Table 3.

Experimentally measured temperatures in the dependence on time for the laser power of 600 W, 800 W and 1000 W are plotted in Figures 3b-d. As it follows from these figures, the maximum measured temperatures are influenced not only by the laser power and welding speed but also very significantly by the distance of thermocouples from the welding centerline. The maximum temperatures on the level of 602 °C and 735 °C were recorded for the laser power of 600 W and the welding speed of 10 mm.s^{-1} and the laser power of 1000 W and the welding speed of 14 mm.s^{-1}, respectively, with the thermocouples located the closest to the centre of the weld joint. Due to the more distant location of thermocouples during the laser welding with higher welding speeds and the same laser power of 800 W, the temperatures measured during the welding with the welding speed of

10 mm.s^{-1} are lower than that measured by the welding with the welding speed of 17 mm.s^{-1} and also 14 mm.s^{-1} (Figure 3c).

Table 3. Welding parameters and position of thermocouples in performed experiments.

Sample	*w* [mm.s^{-1}]	*P* [W]	Distance of thermocouples TC1 and TC2 from the welding line	
			a [mm]	*b* [mm]
1	10	400	6.85	6.36
2	10	600	1.30	2.11
3	10	800	6.83	6.27
4	14	600	2.53	3.75
5	14	800	3.48	4.33
6	14	1000	1.46	1.36
7	17	600	3.72	3.56
8	17	800	2.31	2.61
9	17	1000	6.17	6.48

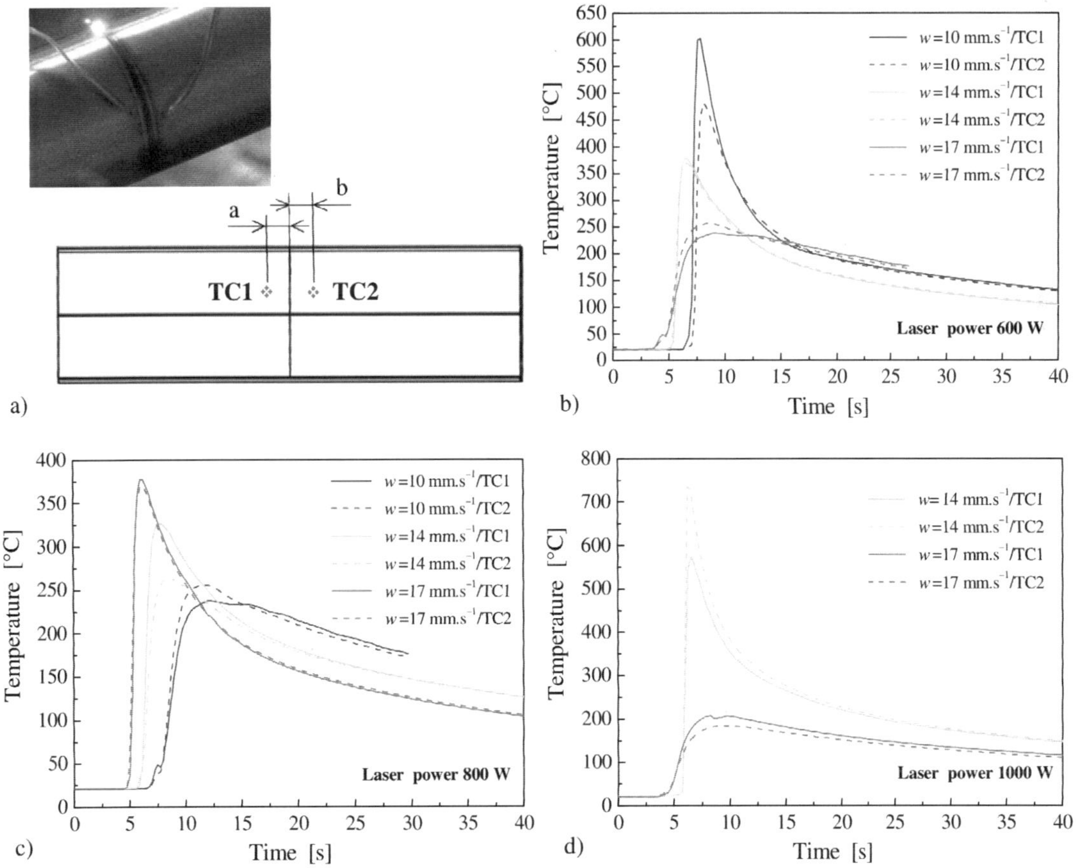

Figure 3. Schematic illustration of the thermocouple positioning (a) and the time dependence of measured temperatures with the laser power of a) 600 W, b) 800 W and c) 1000 W using different welding speeds.

In Figure 4, the measured and computed time histories of temperatures during the laser welding with chosen parameters are compared together with the appearance of produced weld joints. The computed temperatures correspond to the results obtained in the node located at the distance „*a*" indicating in the graphs, i. e. in the node nearest to the position of the thermocouple TC1 or TC2. For the laser power of 600 W and the velocities of 10 mm.s^{-1} and 17 mm.s^{-1} (Figure 4a, c), the distance of corresponding nodes from the weld centerline is smaller than the distance of thermocouple what results in higher maximal computed temperatures in comparison with the measurements. On the other hand, in the case illustrated in Figure 4b (P = 1000 W and w = 17 mm.s^{-1}), the thermocouple TC1 is located closer to the centre of the weld joint than the corresponding node, so the maximum temperature measured by the thermocouple TC1 is higher.

In all investigated cases, the computed temperature decrease during the weld cooling is not as rapid as during the measurements. In simulation model, the weld cooling to the shielding gas and to the environment was neglected and in this reason the computed temperatures during the cooling phase are slightly higher. However, based on the comparison of measured and calculated temperatures it can be concluded that the simulation model was prepared correctly and it can be used for the following computation in order to set proper welding parameters for the laser welding of thin-walled tubes from the AISI 304 stainless steel.

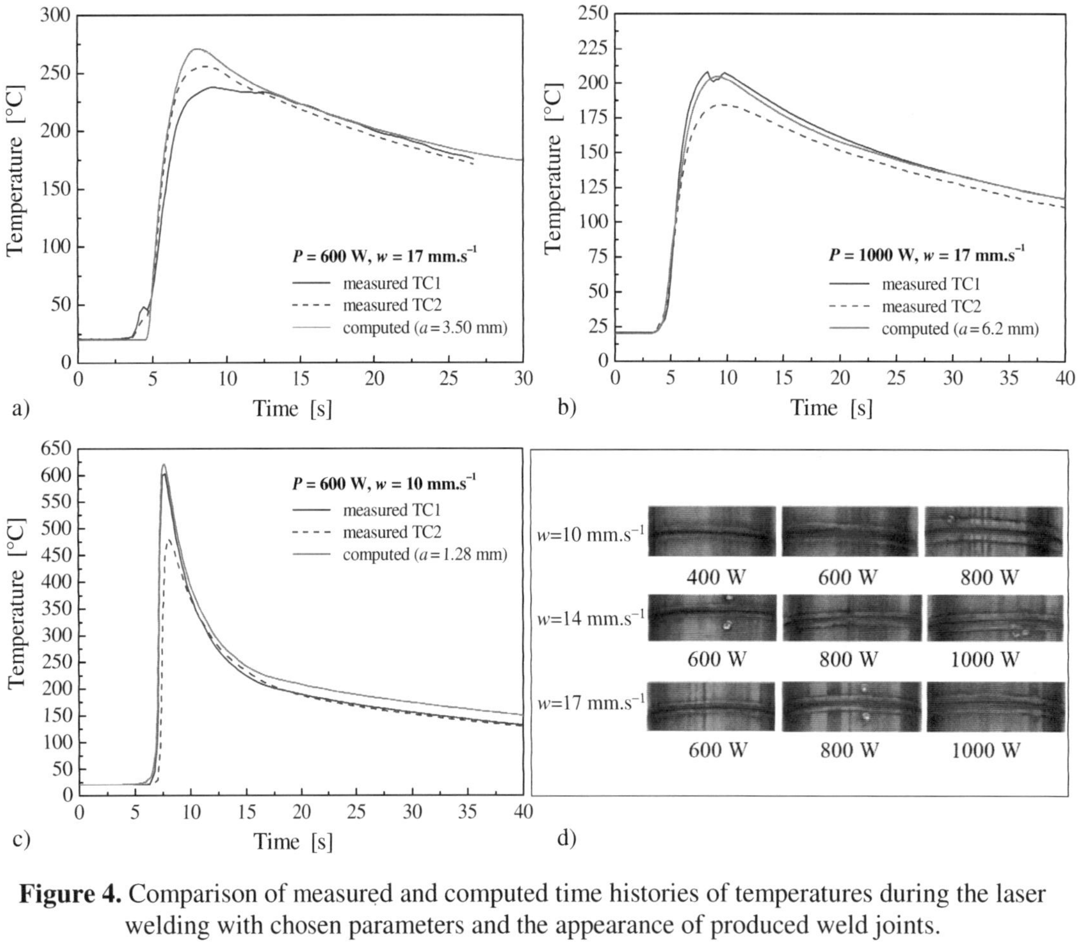

Figure 4. Comparison of measured and computed time histories of temperatures during the laser welding with chosen parameters and the appearance of produced weld joints.

5. Results and discussion

The developed and verified simulation model was applied for the design of welding parameters for the laser welding of the thin-walled tubes from the AISI 304 steel. A set of numerical simulations in the program code ANSYS was carried out with the welding speed varied from the required interval from 30 mm.s^{-1} to 60 mm.s^{-1} and the supposed values of laser power from 500 W to 1200 W.

For illustration, the temperature fields computed applying chosen welding parameters are depicted in Figure 5. The increase in the laser power at the constant welding speed results in the enlargement of a fusion zone and as well as in the enhancement of the maximum temperature of the weld metal (Figure 5a-c). On the other hand, with the increase in welding speed, the size of a fusion zone decreases (Figure 5d-f). The highest temperatures in fusion zone were reached when the combination of high laser power and low welding speed was applied.

In Figure 6, the temperatures of a weld pool and a weld root are shown for considered welding speeds and chosen values of laser power. Based on presented time-temperature dependences, the maximum temperatures attained at the top of the weld pool can be estimated and the possibility of the weld root re-melting can be assessed.

For the welding speed of 30 mm.s^{-1} and the laser power of 500 W, the maximum temperature of the weld root is deeply below the liquidus and also the solidus temperature of the AISI 304 steel. This laser power is insufficient for the production of a sound weld joint. By the laser power of 700 W, the weld root can be re-melted as the maximum weld root temperature reaches the liquidus temperature T_L. In case of welding with the higher laser power (800 W and more), the weld root temperatures are too high.

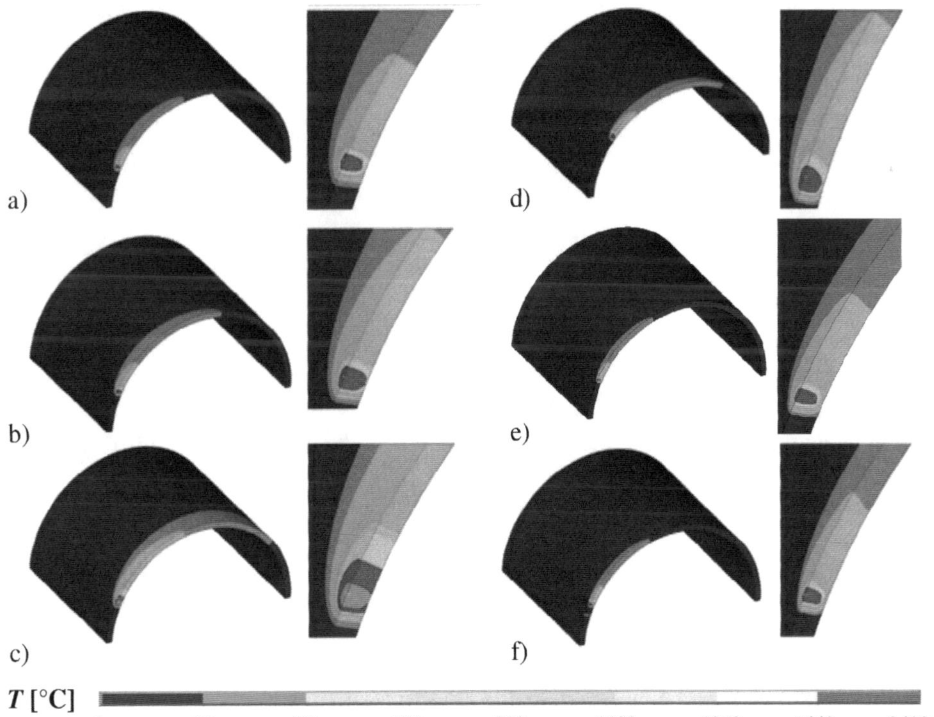

Figure 5. Temperature distribution in a welded tube during the laser welding with the chosen welding parameters: welding speed of 17 mm.s^{-1} and the laser power of a) 500 W, b) 600 W, c) 1000 W and d) the welding speed of 30 mm.s^{-1} and the laser power of 800 W, e) the welding speed of 50 mm.s^{-1} and the laser power of 800 W, f) the welding speed of 60 mm.s^{-1} and the laser power of 900 W.

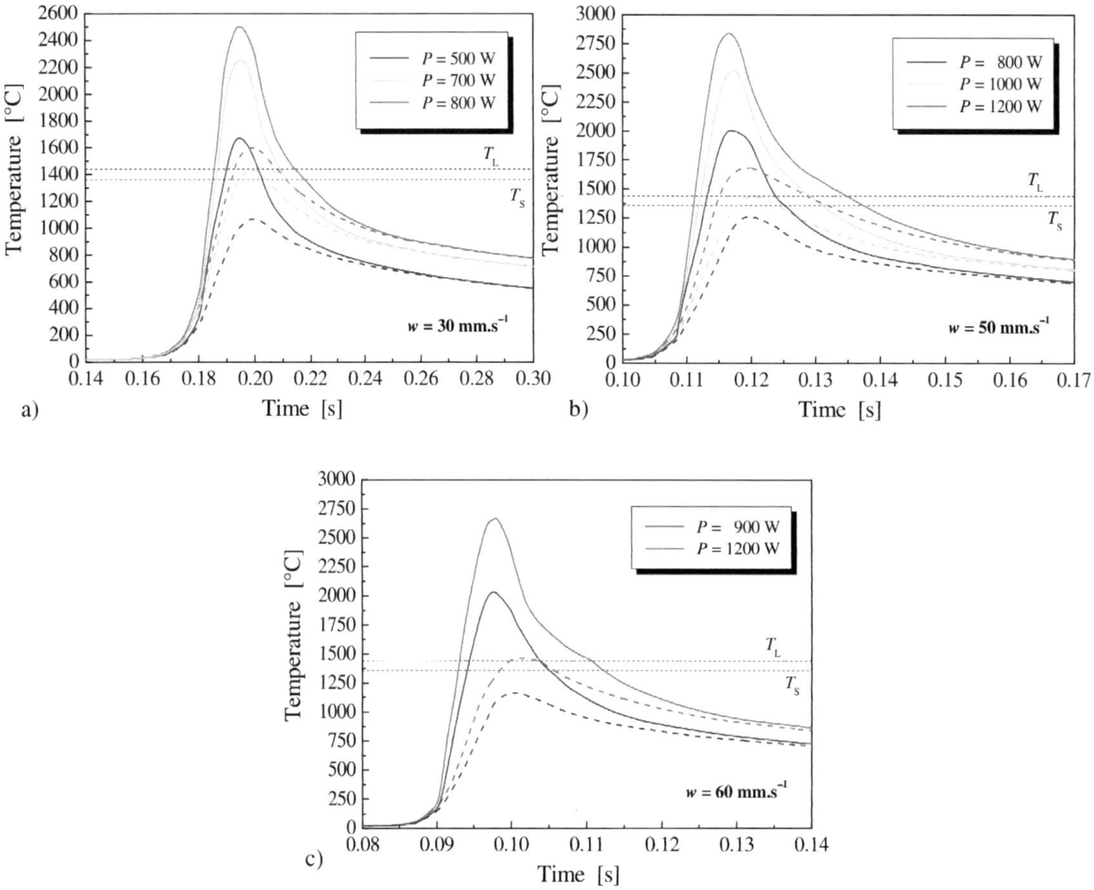

Figure 6. Time dependence of the temperatures at the top of weld pool and on the weld root computed for the welding speed of a) 30 mm.s^{-1}, b) 50 mm.s^{-1} and c) 60 mm.s^{-1} and the chosen values of laser power.

Similarly, the laser power of 800 W and 900 W is not sufficient for the production of high-quality weld joint of tubes from AISI 304 steel if the welding speed is 50 mm.s^{-1} or more. The temperatures of a weld pool using the laser power of 1200 W together with the welding speed of 50 mm.s^{-1} are very high. They can lead to the unwanted boiling of the weld metal.

Based on performed numerical simulations, the welding schedule for laser beam welding of thin-walled tubes from the AISI 304 steel was proposed for the required welding speeds from 30 mm.s^{-1} to 60 mm.s^{-1} (Figure 7). In this figure, the details of temperature distribution in the longitudinal weld cross-section for considered welding parameters are shown. In the grey-coloured regions, the temperatures are above the 2100 °C.

Figure 7 clearly indicates in which cases the tube wall is re-melted through the whole thickness and which welding parameters can be appropriate for the production of high-quality weld joints. For the welding speed of 30 mm.s^{-1}, the minimal laser power must be 700 W. The application of laser powers above 800 W can cause boiling and evaporation of the weld metal. For the welding speed of 50 mm.s^{-1}, the laser power from the interval from 900 W to 1000 W can be recommended. For the highest considered welding speed of 60 mm.s^{-1}, the minimal laser power of 1000 W is required. However, the laser power more than 1100 W can lead to the weld pool temperature increase above the boiling temperature of the AISI 304 steel.

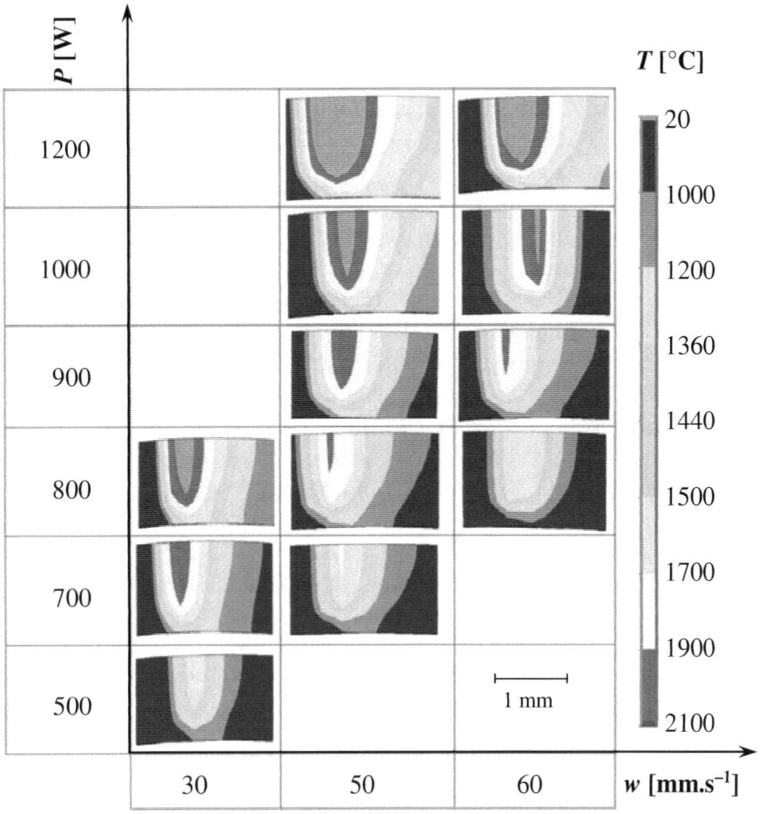

Figure 7. Welding schedule for the laser beam welding of thin-walled tubes from the AISI 304 steel.

6. Conclusions

Based on the developed and verified simulation model, the temperature fields during the process of laser beam welding of thin-walled tubes from the AISI 304 steel were investigated and analysed. According to obtained results, the following values of laser power can be recommended for the production of sound weld joints of thin-walled tubes from AISI 304 stainless steel using the solid-state disc laser TRUMPF TruDisk 4002 and required welding speeds:

- the laser power from 700 W to 800 W for the welding speed of 30 $mm.s^{-1}$,
- the laser power from 900 W to 1100 W for the welding speed of 50 $mm.s^{-1}$ and
- the laser power from 1000 W to 1200 W for the welding speed of 60 $mm.s^{-1}$.

Acknowledgments

The research has been supported by the Scientific Grant Agency of the Slovak Republic (VEGA) within the project No. 1/1010/16 and the Programs to support excellent young researchers and research teams (NUM-LBW-FE/Al and 1347/PRO-MAT-ZVAR) of the STU Bratislava.

References

[1] Katayama S 2013 *Handbook of Laser Welding Technologies*. A volume in Woodhead Publishing Series in Electronic and Optical Materials (Cambridge: Woodhead Publishing Limited)

[2] Lawrence J 2008 *Advances in Laser Materials Processing, Technology, Research and Application*. A volume in Woodhead Publishing Series in Welding and Other Joining Technologies (Cambridge: Woodhead Publishing Limited)

[3] Michalos G et al. 2010 Automotive assembly technologies review: challenges and outlook for a flexible and adaptive approach *CIRP-JMST* **2** 81-91
[4] Rossini M et al. 2015 Investigation on dissimilar laser welding of advanced high strength steel sheets for the automotive industry *Mat Sci Eng A-Struct* **628** 288-296
[5] Hong K Yand Shin C 2017 Prospects of laser welding technology in the automotive industry: A review *J Mater Process Tech* **245** 46-69
[6] Choudhury B and Chandrasekaran M 2017 Investigation on welding characteristics of aerospace materials – A review *Materials Today: Proceedings* **4** 8 7519-7526
[7] Blackburn J 2012 *Laser welding of metals for aerospace and other applications* A volume in Woodhead Publishing Series in Welding and Other Joining Technologies (Cambridge: Woodhead Publishing Limited) pp 75–108
[8] Reitemeyer D, Schultz V, Syassen F, Seefeld T and Vollertsen F 2013 Laser Welding of Large Scale Stainless Steel Aircraft Structures *Physics Proc* **41** 106-111
[9] Moraitis G A and Labeas G N 2009 Prediction of residual stresses and distortions due to laser beam welding of butt joints in pressure vessels *Int J Pres Ves Pip* **86** 2–3 133-142
[10] Zhou Y N and Breyen Mark D 2013 *Joining and Assembly of Medical Materials and Devices.* A volume in Woodhead Publishing Series in Biomaterials (Cambridge: Woodhead Publishing Limited)
[11] Dal M and Fabbro R 2016 An overview of the state of art in laser welding simulation *Opt Laser Technol* A **78** 2-14
[12] Chukkan R J et al 2015 Simulation of laser butt welding of AISI 316L stainless steel sheet using various heat sources and experimental validation *J Mat Process Techn* **219** 48-59
[13] Mackwood P and Craferb R C 2004 Thermal modelling of laser welding and related processes: a literature review *Opt Laser Technol* **37** 99
[14] Frewin M R and Scott D A 1999 Finite Element Model of Pulsed Laser Welding *Weld J* **78** 5-22
[15] Lindgren L E 2004 Numerical modelling of welding *Comput Methods Appl Mech Eng* **195** 48-49 6710-6736
[16] Olabi A G and Gasalino G 2014 Mathematical modeling of weld phenomena part I Finite element modeling *Reference Module in Materials Science and Materials Engineering, Comprehensive Materials Processing* **6** 101-109
[17] Nagel F, Simon F, Kümmel B, Bergmann J P and Hildebrand J 2014 Optimization Strategies for Laser Welding High Alloy Steel Sheets *Physics Proc* **56** 1242-1251
[18] Sokolov M, Salminen A, Khlusova E I, Pronin M M, Golubeva M and Kuznetsov M 2015 Testing of New Materials and Computer Aided Optimization of Laser Beam Welding of High-Strength Steels *Physics Proc* **78** 255-264
[19] Ashby M F and Smidman M 2010 *Materials for Nuclear Power Systems* (Granta material inspiration) 1-20
[20] Xu J et al 2017 Comparison of residual stress induced by TIG and LBW in girth weld of AISI 304 stainless steel pipes *J Mat Process Techn* **248** 178-184
[21] https://www.trumpf.com/en_GB/products/laser/disk-lasers/trudisk/
[22] ANSYS Inc. Computer software Pennsylvania, United States
[23] Lindgren L E 2001 Finite element modelling and simulation of welding *J Ther Stresses* **24** 141-192
[24] Souloumiac B, Boitout F and Bergheau J-M 2002 A new local global approach for the modelling of welded steel component distortions, *Math Modelling Ser* **6** 573–590
[25] Deng D 2009 FEM prediction of welding residual stress and distortion in carbon steel considering phase transformation effects *Mater Design* **30** 2 359-366
[26] Incropera P F and De Witt D P 1996 *Fundamentals of heat and mass transfer* (New York: J. Wiley and Sons)
[27] JMatPro Help, Release 6.1 2012 Sente Software Ltd
[28] Huy H and Argyropoulos S A 1996 Mathematical modelling of solidification and melting: a review *Model Simul Mater Sc* **4** 371–396

IOP Conf. Series: Materials Science and Engineering **266** (2017) 012013 doi:10.1088/1757-899X/266/1/012013

[29] Yau Ch L Chao L S 2006 Modified Effective Specific Heat Method of Solidification Problems *Mater Trans* **47** 2737-2744
[30] Bergström D 2008 *The Absorption of Laser Light by Rough Metal Surfaces* - Doctoral Thesis (Luleå: Luleå University of Technology, Sweden)

IOP Conf. Series: Materials Science and Engineering **266** (2017) 012014 doi:10.1088/1757-899X/266/1/012014

The influence of additives on crystallization of blends based on polylactid acid

D Perďochová[1], K Tomanová[1], P Alexy[1], J Bočkaj[1], J Feranc[1], R Plavec[1], L Omaníková[1], P Jurkovič[1] and R Přikryl[2]

[1]Slovak University of Technology, Faculty of Chemical and Food Technology, Institute of Natural and Synthetic Polymers, Department of Plastics, Rubber and Fibers, Radlinského 9, 812 37 Bratislava, Slovakia,
[2]Brno University of Technology, Institute of Material Chemistry, Faculty of Chemistry, Purkyňova 118, 612 00 Brno, Czech Republic

dagmaraperdochova@gmail.com

Abstract. The sustainable development consists of the effort of replacing commonly used polymer materials for the biodegradable ones. They do not have sufficient physical and mechanical properties, therefore they have to be modified by producing various ratio mixtures or with the additives. Improving their processability and properties is an important challenge to be afforded before using these materials on the market. One way to improve the properties of these materials is to prepare their blends [1]. The most common way of the preparation of packaging materials is the injection moulding, where the crystallization of material is very important. That is the reason the crystallization has become one of the most studied characteristics of biodegradable blends based on PLA / PHB. The work is a contribution to works that deal with the description of the structure of PLA / PHB blends, which have been modified by the addition of various types of additives, mainly nucleating agents and plasticizers [2]. The films of these blends keep approximately unchanged mechanical properties after two weeks of storage. The presence of plasticizer (ATBC) and nucleating agent (BN) has no significant impact on the processing stability of PLA/PHB blends. The results show that the combination of the plasticizer and nucleating agent greatly affects the process of crystallization of PLA/PHB blends as well as the ratio of polymers and the heat treatment process.

1. Introduction

Nowadays, the most commonly used materials are polymers, both synthetic and natural. Their expansion is based on their good physical properties, mechanical properties, and also because of low production costs. The increasing consumption of conventional synthetic polymeric materials is the cause of the growing problem of their disposal. At the same time, we are faced with an ecological problem. For this reason, efforts are being made to replace conventional plastics (PE, PP, PS and others) with biodegradable plastics and therefore more environmentally friendly. One possibility how to solve this problem can be by the production of biodegradable plastics. Biodegradable polymers, especially those from renewable resources and their blends, are playing an important role in the field of research of new materials. Of all of the potential products made from annually renewable resources, polylactic acid (PLA) is currently the most interesting. This biopolymer is 100% renewable, it can be

Content from this work may be used under the terms of the Creative Commons Attribution 3.0 licence. Any further distribution of this work must maintain attribution to the author(s) and the title of the work, journal citation and DOI.
Published under licence by IOP Publishing Ltd

easily composted and has comparable properties to commercially-known plastics [3,4]. The second group of studied polymers are polyhydroxyalkanoates, biopolymers synthesized by the action of microorganisms. As in the case of polylactic acid, these are polymers with similar properties to conventional plastics, they are completely biodegradable and have potential as a biodegradable substitute. The simplest and most commonly used poly-β-hydroxyalkanoate is poly-β-hydroxybutyrate (PHB) [5]. Polymers that are produced from natural resources and are capable of being decomposed in nature have worse physical and mechanical properties than conventional polymer materials and therefore, use of them in demanding applications and to replace conventional plastics, they have to be modified in some ways. One way to modify the properties of these materials is to prepare their blends. This method of the preparation can achieve a synergistic improvement in properties in the comparison with the pure components [6]. The properties of polymeric materials are generally improved during processing in accordance with the requirements of the final product application. One important factor for setting the processing conditions is the understanding and description of the polymer structure, which affects the final properties of the product. Based on the supramolecular structure, we can predict and control the resulting behaviour of polymeric materials [7]. Polymer blends are prepared as a result of improved processing and utility properties together with various additives and modifiers. The plasticizers are one of the most commonly applied additives to improve the processability and flexibility of the final product. Since the most common method of preparing polymeric materials is injection moulding where crystallization of the material is a very important factor, crystallization has become one of the most studied characteristics of PLA / PHB blends [8]. At the same time the crystallization is followed by the effect of other additives, nucleating agents on the crystallization process itself and their effect on the resulting mechanical properties. The present work is a contribution to the number of works dealing with the description of the crystalline structure of PLA / PHB blends which have been modified by the addition of various types of additives, mainly nucleating agents and plasticizers.

2. Materials

Polylactid acid $(C_3H_4O_2)_n$ – Ingeo 4043D, Mw = 202 kDa, PDI = 2,18 ; D-isomer content 4,2 %. Nature Works, USA
Polyhydroxybutyrate $(C_4H_6O_2)_n$, Mw = 608 kDa, PDI = 4,23; homopolymer P3HB - Biomer, Nemecko
Ester-type plasticizer – acetyltributylcitrate (ATBC), Jungbunzlauer, Nemecko (0, 5, 10, 15, 20, 25%wt.)
Nucleating agent – Boron Nitride (BN), Momentive, USA (BN – 0, 1, 5 % wt.)

3. Experimental

3.1 Blends preparation

Biodegradable blends based on PLA/PHB (60/40 w/w) with the specific content of additives were prepared using twin screw extruder (L/D = 40, D = 16 mm) at temperature 160 – 170 – 180 – 180 – 185 – 185 – 185 – 180 – 175 – 170 °C (direction hopper – head). Extruded materials were chilled in water and granulated into small pellets. Pellets were dried at 70°C for 2 hours in hot air oven and consequently used for rheological and DSC measurements.

3.2 Processing stability

The oscillation rheometer RPA 2000 was used for processing stability testing of prepared blends. Timed test was done at three temperatures: 170, 180, 190°C, the angle of strain was 30° and oscillation frequency was 60 cpm. For correct evaluation of temperature influence on the rate of polymer degradation based on the decrease of complex viscosity, the elimination of differences in initial viscosity was done by recalculation of complex viscosity using the formula (1):

$$\eta^*_{rel}(t) = \eta^*(t) / \eta^* \quad (1)$$

where $\eta^*_{rel}(t)$ is relative complex viscosity at time t, $\eta^*(t)$ is complex viscosity at time t and η^* is complex viscosity at the start of the test.

3.3 Films preparation

For films preparation was used single-screw Brabender extruder (screw diameter 19 mm, L/D=25) as well as chill-roll technology was used. Prepared films were used for measuring the mechanical properties. The selected temperature profile of the single screw extruder in the head → hopper direction was as follows: 170-180-180-170 ° C.

3.4 Mechanical properties measurement

Tensile strength at break (σb) and the elongation at break (εb) were determined according to ISO 527 standard using Zwick machine at cross-head speed 50 mm/min. Strips 15mm width were used as testing pieces.

3.5 DSC measurement

DSC (Mettler Toledo) was used for studying the crystallization properties of prepared blends under specific conditions. The flow of purge gas of nitrogen was 50ml/min. Standard aluminium crucibles were used. Measuring program:

1. Heating 10 K/min, 30-190 °C
2. Isothermal process 190 °C, 3 min
3. Cooling 10 K/min, 190-30 °C
4. Isothermal process 30 °C, 3 min
5. Heating 10 K/min, 30-190 °C

4. Results and discussion

The work deals with the study of influence of plasticizer and nucleating agent on crystallization process and the resulting thermal and physical properties of PLA/PHB blends.

4.1 Processing stability

The indication of the first stages of degradation process of polymers can be easily detected using the oscillation rheometer. Changes in relative complex viscosity of the prepared blends were taken as the indicator of processing stability. According to article [9], it was shown that changes in relative complex viscosity reflect changes in molecular weight of degraded polymer very well. The example of dependency of complex viscosity and relative complex viscosity on test time for PLA/PHB blends including 1% wt. of nucleating agent and different concentration of plasticizer is shown in Figure 1.

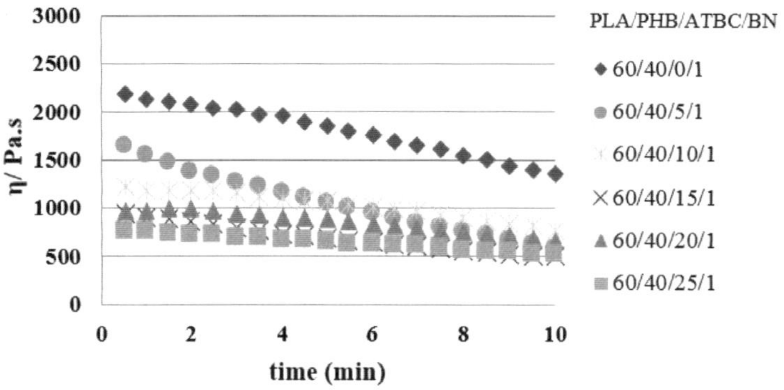

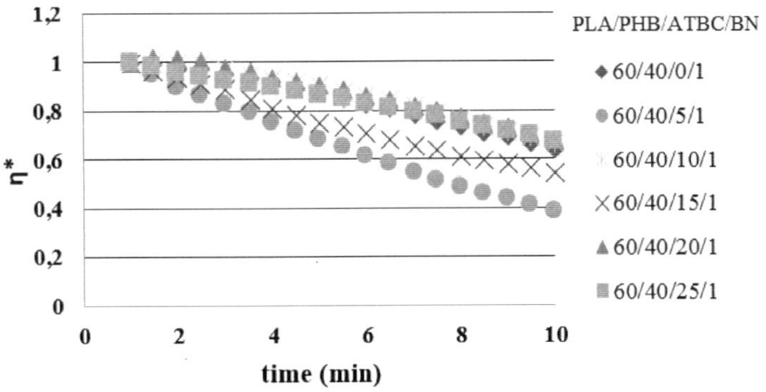

Figure 1. The dependency of complex viscosity and relative complex viscosity on test time for PLA/PHB blends including 1% wt. of nucleating agent and different concentration of plasticizer

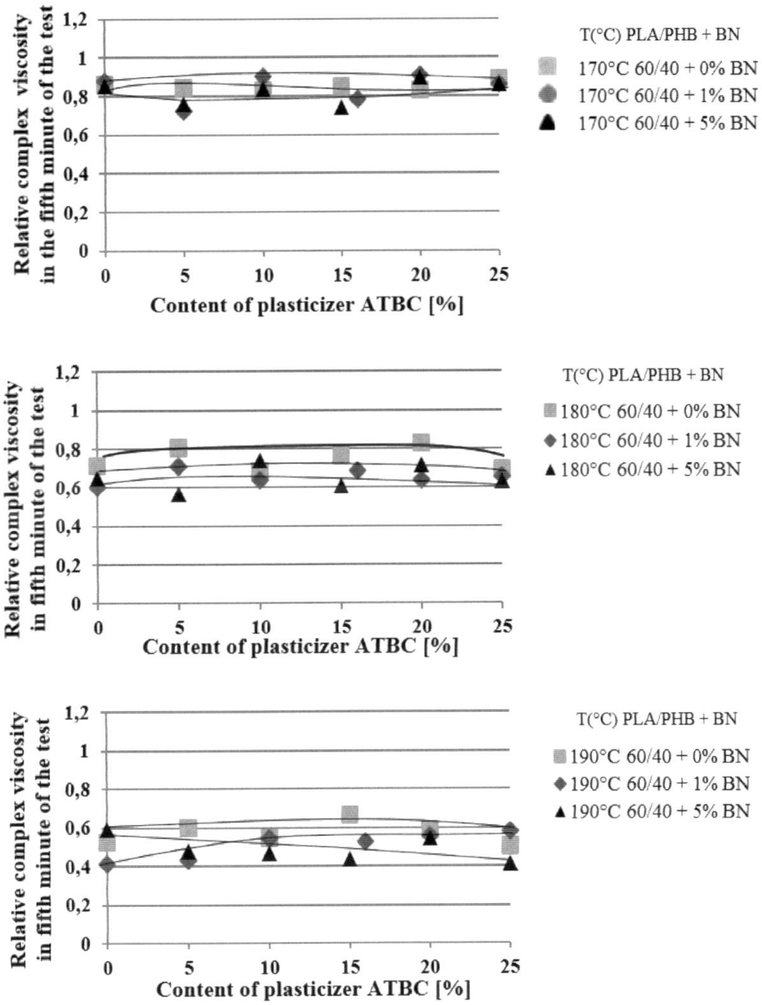

Figure 2. The dependency of relative complex viscosity in the fifth minute of the test on content of plasticizer and nucleating agent BN at different temperatures (170 °C, 180 °C, 190 °C).

According to results shown in Figure 2 it can be seen that the viscosity of the blends in the fifth minute of the test, because of the thermo-mechanical stress caused by oscillatory rheometer, gradually decreases with temperature and it indicates that the increasing temperature causes the expected increase of degradation of the polymer blends. Differences in relative complex viscosity are significant but slightly. The effect of the presence of the nucleating agent, as well as the influence of the concentration of plasticizer is minimal and the observed differences can be considered to be within the experimental variability. Consequently, neither the plasticizer nor the nucleating agent has a negative effect on the processing stability or rheological characteristics of the studied blends.

4.2 Mechanical properties

The PLA / PHB blends were prepared according to method mentioned in chapter 3.1. Subsequently, the physical-mechanical properties of PLA / PHB blends were measured according to method mentioned in chapter 3.4 after sample preparation at time intervals 24 hours, a week and 2 weeks after the extrusion. The conditions for storage of samples were 50% relative humidity and 25°C. Nucleating agent BN is also used as a filler in high performance in polymer blends, which greatly improves thermal conductivity. In film production, this property of BN appears to have been evident. The addition of a larger amount of BN (5 wt.%) has a positive effect on the processing properties either during the production of the pellets or the film, probably due to the better thermal conductivity of the prepared blends. Whether cooling in a water bath or cooling on chill rolls, it is likely that a faster heat transfer occurs, resulting in the stiffness of string and a reduced stickiness of the films on the chill rolls. For all samples (PLA/PHB : 60/40 w/w ratio), physical-mechanical properties were maintained after 2 weeks of storage. For this reason, in the following, we only summarize the PLA / PHB blends, depending on the different plasticizer and nucleating agent content after 2 weeks of storage. The elongation at break for 60/40 w/w PLA / PHB blends (Figure 3) with 0, 1 and 5 wt.% of nucleating agent is noted after the addition of 15 %wt. of the plasticizer, and above this plasticizer concentration no change are observed in the elongation with the increase amount of the plasticizer. The addition of the nucleating agent does not affect the elongation at break and value of ε_b can to be considered as similar except the highest concentration of both additives when the elongation at break is slightly lower.

The tensile strength at break (Figure 4) decreases with the addition of 0-15% of plasticizer, the most significant decreasing can be observed in the range of 10-15%, from 15% of plasticizer we do not notice a significant change in tensile strength at break. The differences in tensile strength at break caused by presence of nucleating agent are more significant in case of low concentration of plasticizer to compare with differences at higher concentration of plasticizer, which are very small.

Based on described mechanical properties of prepared blends the marginal effect of presence of nucleating agent was observed in case of both tensile parameters. More significant effect was noticed in the case of presence of plasticizer as additive in different concentrations.

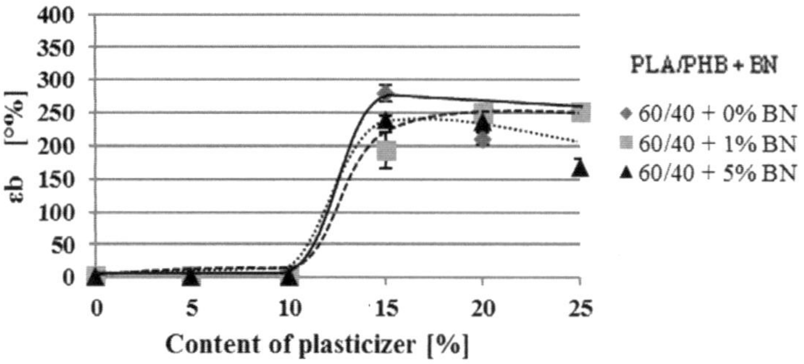

Figure 3. The dependency of elongation at break on content of plasticizer after two weeks of storage.

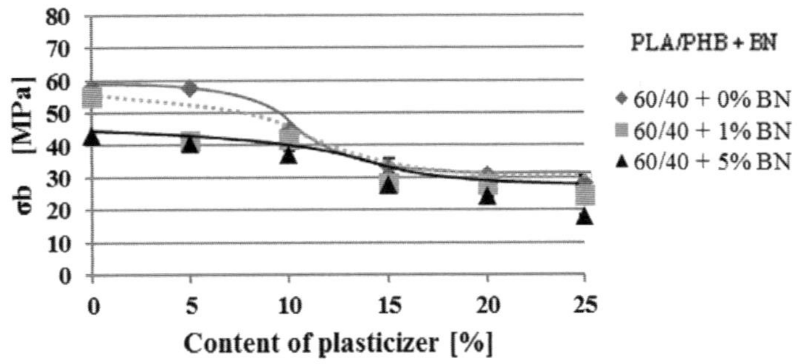

Figure 4. The dependency of tensile strength at break on content of plasticizer after two weeks of storage.

4.3 Thermal properties

The thermal properties of prepared blends were studied by DSC, especially the crystallization process. It is well known that the first heating curve is significantly influenced by the sample´s thermal history. The thermal history of prepared samples was erased by first heating run which is not reported because it reflects more thermal history of the sample preparation than real material characteristics. Only cooling and the second heating curves are presented in Figure 5 and Figure 7. The curves of cooling are shown in Figure 5, dependency of crystallization temperature Tc (peak) on nucleation agent as well as plasticizer concentration is shown in Figure 6. It can be seen that the addition of nucleating agent significantly influences the crystallization of PLA/PHB blends. The crystallization peak is shifted to a higher temperature in comparison to blank PLA/PHB blend. The changes of Tc depend also on plasticizer concentration. With increasing content of plasticizer, crystallization peak of nucleated blends is shifted to lower temperature. It means, that during the cooling of the melt, the crystallization process starts earlier as an effect of nucleation agent and it starts later as an effect of plasticizer, probably because plasticizer allows better mobility of the polymer chains. It can be predicted that the presence of other types of plasticizers can exhibit more or less strong effect on crystallization process of PLA/PHB blends in presence of nucleating agent in dependency of interaction between plasticizers and polymers. It means that by suitable selection of nucleation agent and plasticizer as additive in combination, the temperature of crystallization can be modified.

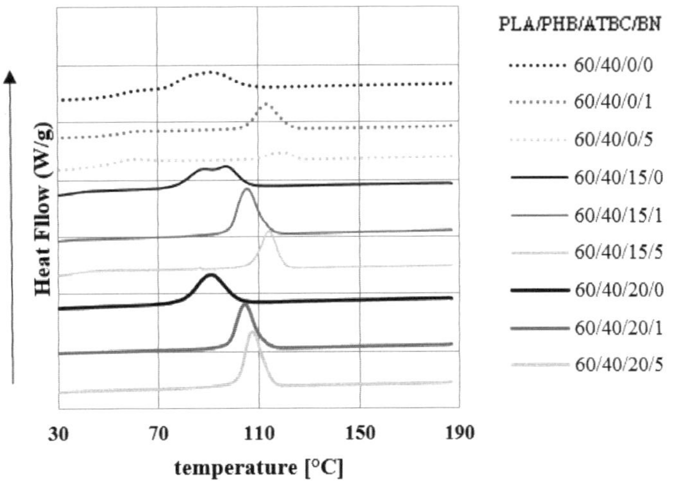

Figure 5 The curves of cooling of prepared blends. Ratio: PLA/PHB/plasticizer/nucleating agent.

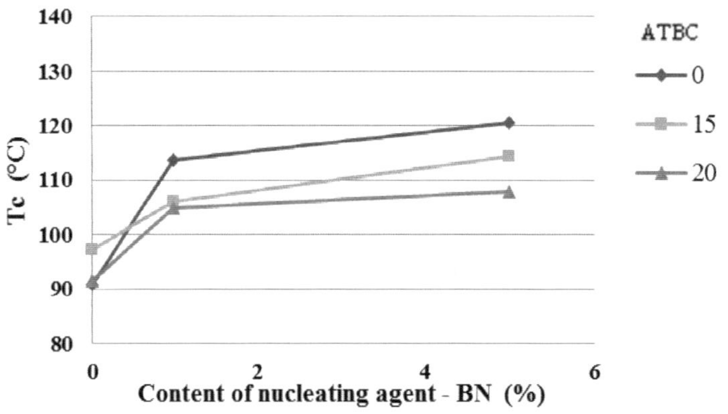

Figure 6 The dependency of Tc on nucleation agent content for various concentration level of plasticizer.

The curves of the second heating are shown in Figure 7, where three processes can be observed - the cold crystallisation, the melting of PLA and the melting of PHB. The effect of cold crystallization is present due to low rate of crystallization of PLA during cooling of samples. The dependency of Tcc as the peak value on both additives concentration is shown in Figure 8. The difference in nucleating activity is observed according to presence of nucleating agent as well as plasticizer. The Tcc of cold crystallization is shifted to higher temperature with increasing content of nucleating agent, mainly if blend contains no plasticizer. On the other hand, the increasing content of plasticizer shifts Tcc to lower temperature. This effect allows to modify crystallization process via the processing condition during the processing (temperature of the mould, in injection moulding process for example). Figure 9 shows the dependencies of Tm for PLA as well as for PHB on both additives concentration. Both melting temperatures (for PLA and PHB polymers in the blend) are strongly shifted to lower values by plasticizer concentration. The activity of nucleating agent is influenced by plasticizer as well. It can be seen, based on DSC results, that the crystallization of the blends and the type of crystals formed can be significantly influenced by the influence of the additives.

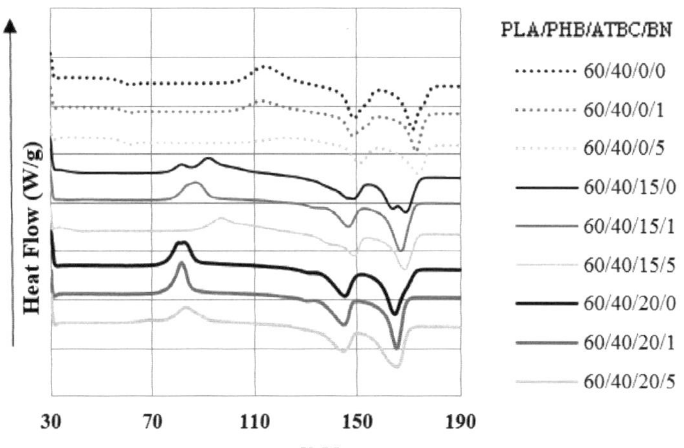

Figure 7. The curves of the second heating of prepared blends. Ratio: PLA/PHB/plasticizer/nucleating agent.

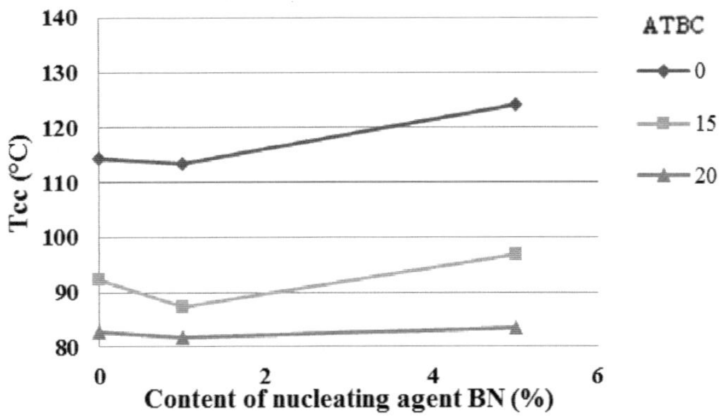

Figure 8. The dependency of Tcc on nucleating agent concentration for various levels of plasticizer content.

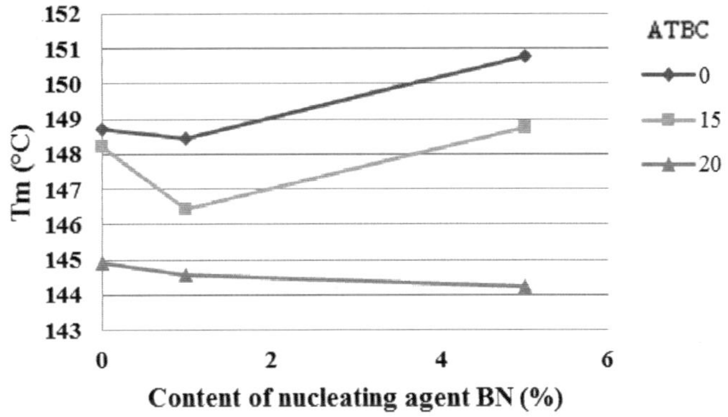

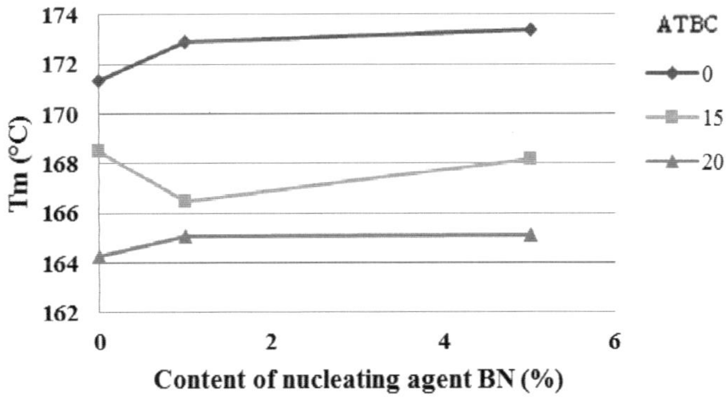

Figure 9. The dependency of Tm on nucleating agent concentration for various levels of plasticizer content.

Polymorphism, but also the process of cold crystallization and melting, is significantly affected. Generally, the nucleating agent (BN) increases the rate of crystallization process of PLA / PHB blends and the plasticizer improves the mobility of the segments to shift the temperature of melting and cold crystallization to lower temperature during the heating of the blend.

5. Conclusion

Neither the plasticizer nor the presence of a nucleating agent has significant impact on the processing stability of PLA / PHB blends. The addition of the nucleating agent has negative impact on the physical-mechanical properties of prepared blends as compared to blends without nucleating agent, especially in the case of tensile strength at break (σb) which decreases compared to samples without nucleating agent. The elongation at break significantly improves over 10% of plasticizer content. Additives, in our case nucleating agent and plasticizer, affect the crystallization process. In general, the nucleating agent (BN) improves nucleating activity of PLA/PHB blends and plasticizer allows better mobility of the chains that results in lower crystallization temperature. The next study will be based on optimizing the composition and processing conditions of PLA/PHB blends.

Acknowledgements

This work was supported by the Slovak Research and Development Agency (Project No. APVV-14-0301 and No. APVV-14-0175)

References

[1] Armentano I, Fortunati E and Burgos N, 2015 Processing and characterization of plasticized PLA/PHB blends for biodegradable multiphase systems *Polymer Letters* **9** 583–596
[2] Lim J S, Park K I and Chung G S 2013 Effect of composition ratio on the thermal and physical properties of semicrystalline PLA/PHB-HHx composites *Materials Science and Engineering* 2131–2137
[3] Gautam R, Bassi A S and Yanful EK 2007 A Review of Biodegradation of Synthetic Plastic and Foams *Appl Biochem Biotechnol* **141** 85-108
[4] Steinbuchel A and Merchessault R H 2005 Biopolymers for medical and pharmaceutical applications. *Polylactides* ISBN 978-3-527-311 186-219
[5] Ansari S and Fatma T. 2014 Polyhydroxybutyrate - a Biodegradable Plastic and its Various Formulations *International Journal of Innovative Research in Science, Engineering and Technology* **3** 9494–9499
[6] Abdelwahab M A and Flynn A 2012 Thermal, mechanical and morphological characterization of plasticized PLA–PHB blends. *Polymer Degradation and Stability* **97** 1822-1828
[7] Gerard T 2014 Polylactide/poly(hydroxybutyrate-co-hydroxyvalerate) blends: Morphology and mechanical properties. *Express Polymer Letters* **8** 609-617
[8] Kaufman S H and Falcetta J J 1977 Introduction to Polymer Science and Technology *Wiley* NY 13-23
[9] Mikušová M, Mihalík M, Alexy P, Tomanová K, Plavec R, Bočkaj J and Vanovčanová Z 2014 Method for testing of processing stability of biodegradable polyesters based on oscillation rheometry. *KGK-Kautschuk Gummi Kunststoffe* ISSN 0848-3276 51-54

IOP Conf. Series: Materials Science and Engineering **266** (2017) 012015 doi:10.1088/1757-899X/266/1/012015

Analysis of the machine frame stiffness using numerical simulation

Š Vrtiel, Š Hajdu and M Behúlová

Slovak University of Technology in Bratislava, Faculty of Materials Science and Technology in Trnava, Ulica Jána Bottu č. 2781/25, 917 24 Trnava, Slovakia

stefan.vrtiel@stuba.sk

Abstract. The paper deals with the numerical modelling and simulation of loads arising during the milling process and their impact on the deformations and stresses developed in the frame of construction of hobby milling machine. Based on the prepared simulation model, the loads during the machining with feed rates in different directions were considered in order to find out the critical loading and its influence on the machine frame deformation. The impact of the frame deformations on the displacement of a tip of the tool, and thus the accuracy of the machining process was evaluated. Results of initial numerical simulations provided also the information about the components of the construction that need to be modified during the optimization of the designed construction and/or the development of a new prototype.

1. Introduction

Compared to the past, the advantages of numerical simulations are increasingly used in the design and development of machine components. While they were once applied mostly for the verification of real experiments, today they are used to the same extent in the phase of a product development [1, 2]. This is confirmed also by the integration of tools for numerical simulation directly to the current CAD systems such as Autodesk Inventor®, SolidWorks®, CREO® CATIA® [3-6] or new interface modules for engineers like ANSYS AIM [7].

Based on the results of numerical analyses and simulations, the new product can be optimized very efficiently. Consequently, duration of the development phase can be significantly shorter. Such a progressive approach to the development process contributes to the increase in competitiveness of an enterprise through the saving time and financial resources [8, 9]. Moreover, application of numerical testing and simulation provide the possibility to investigate several design virtual prototypes of the developed product including geometrical or material options and to evaluate their influence on product behaviour under different boundary conditions and/or loading and in some cases also to predict the efficiency and the product lifetime [10, 11]. These information and knowledge allow the designer to change the product design or even the product concept itself. In some cases, these procedures lead to the discovery of completely new technical solutions. Of course the final product design could not be realized without verification of simulation model and computed results on its physical prototype.

This paper presents the simulation model developed for the initial static stress-strain analysis of the construction frame of a designed hobby milling machine subjected to the different types of loading during a milling process. The main aim of numerical simulations is to evaluate the stiffness of the machine frame and to determine the critical loads causing the maximal deformations and stresses in order to reveal the weak points of construction design.

Content from this work may be used under the terms of the Creative Commons Attribution 3.0 licence. Any further distribution of this work must maintain attribution to the author(s) and the title of the work, journal citation and DOI.
Published under licence by IOP Publishing Ltd

2. Problem description and calculation of a cutting force

The portal 3D milling machine (Figure 1) designed mainly for the hobby segment was subjected to the analysis. The design of the construction took into account the lowest weight of the machine [12] what is related to the load and the required power for the driving units. The basic dimensional characteristics of the machine are summarized in Table 1.

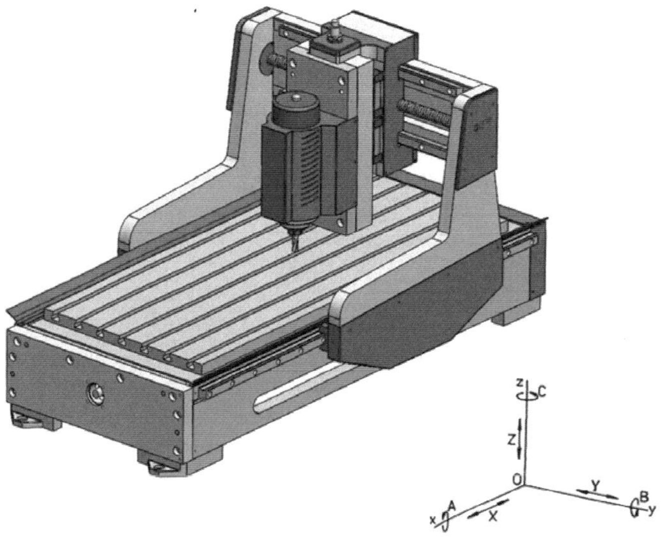

Figure 1. Design of a milling machine.

Table 1. Dimensional parameters of the milling machine.

Machine dimensions	Value	Units	Feed axes (travel)	Value	Units
Length	975	mm	*X*	500	mm
Width	539	mm	*Y*	300	mm
Height	563	mm	*Z*	120	mm

For the simulation, the milling process with a two flute centercutting milling tool was chosen. In this process, the maximum cross-sectional area of the layer to be removed and hence the maximum load is known on the base of the cutting depth a_p, the feed rate f_z and the cutting resistance of the machined material k_c. Considering the maximum cross-sectional area of the removed layer and other selected technological parameters presented in Table 2, the values of the cutting force components were determined using the relationships (2)-(4).

Table 2. Selected technological parameters.

Parameter	**Nomenclature**	**Value**	**Units**
Tool diameter	D	5	mm
Number of tool flute	n_z	2	-
Cutting depth	a_p	2	mm
Cutting width	a_e	5	mm
Cutting speed	V_c	300	m.min^{-1}
Feed rate	f_z	0.1	mm.tooth^{-1}
Angle of heel edge	λ_s	25	degree
Specific cutting resistance	k_{c1}	830	MPa

Calculation of the cutting resistance:

$$k_c = \frac{k_{c1}}{f_z^m} = \frac{830}{0.1^{0.23}} = 1409.54\text{ MPa} \tag{1}$$

The exponent m denotes the slope of the linear dependence of k_{c1} on the chip thickness. Its values for various materials can be obtained on base of experiments [13].

Calculation of the main (tangential) cutting force F_c:

$$F_c = a_p f_z k_c K_\gamma K_{wear} = 2\times 0.1\times 1409.54\times 0.86\times 1.5 = 363.66\text{ N} \tag{2}$$

where K_γ is the front angle correction parameter, K_{wear} is the correction parameter according to wear [13].

Calculation of radial cutting force F_{rad}:

$$F_{rad} = F_c k = 363.66\times 0.7 = 254.56\text{N} \tag{3}$$

where k represents the ratio between radial and tangential cutting force.

Calculation of axial cutting force F_{ax}:

$$F_{ax} = 0.28 F_c tg(\lambda_S) = 0.28\times 363.66\times tg(25°) = 47.48\text{ N} \tag{4}$$

2.1. Analysis of stiffness in single motion modes

In Figure 2, the feed directions V_{f1}, V_{f2} and V_{f3} are illustrated for the possible situations which can occur during milling with given technological conditions. The movement of the tool in the feed directions V_{f1} and V_{f2} is accomplished by moving of the entire machine portal. The V_{f3} feed movement is performed only by the support moving along the portal. Thus, the directions of cutting forces will vary together with the simulated feed direction.

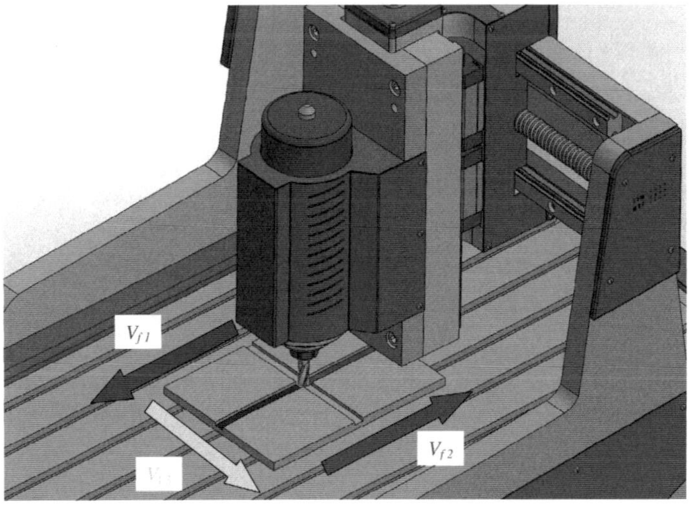

Figure 2. Illustration of possible feed directions V_{f1}, V_{f2} and V_{f3}.

2.2. Simulation model

In order to evaluate the tool stiffness using numerical simulation, the simulation model was developed. The geometrical model was adjusted by the top-down method of modelling [8]. For calculation, the detailed input CAD model was simplified by removal or editing some technological elements which are supposed to have no significant influence on the results of numerical simulation. These changes were also necessary to make to generate appropriate finite-element mesh. Elements such small size threads (M5), small bumps for easier assembly and complex shape areas were removed or adapted. Editing the resulting assembly also consisted of removing the fasteners such as bolts and locking pins. Screws and locking pins were replaced by fixation bonds representing their functionality in the simulation model.

For more accurate results, the option "No Separation" was chosen for the optimal definition of contact on the adjacent areas. This contact description represents the real contact of two adjacent surfaces with the possibility to evaluate the surface pressure in the contact area. On the other hand, it does not remove the degrees of freedom relative to the adjacent component as e. g. the "Bonded" option that prevents them from the movement relative to each other like in the case of the spike contact region of the ball screw and the nut. Overall, thirty-nine contacts were defined in the simulation model.

2.3. Finite element mesh

The accuracy of numerical calculations depends significantly on the density and the quality of generated finite-element mesh. In this reason, remarkable attention was paid to the mesh generation and to its quality evaluation. The final finite-element mesh (Figure 3) contains the total number of 424 238 elements with the maximum element side length of 3 mm.

Figure 3. Model with generated finite element mesh.

2.4. Material model

The material model contains the properties of two materials, EN AW 7075 aluminium alloy and the EN S195T steel given in Table 3. The supporting part of the structure is supposed to be produced from the aluminium alloy, linear guides, ball screws and mounting screws are designed from the steel. Material properties from the database of ANSYS Workbench [14] were applied.

Table 3. Physical and mechanical properties of applied materials [14].

Property	Aluminium alloy	Structural steel	Units
Density	2770	7850	kg.m^{-3}
Young modulus	71000	200000	MPa
Yield strength	280	250	MPa
Tensile strength	320	490	MPa

2.5. Loads and boundary conditions

The distribution of cutting forces varies according to the machining direction (Figure 4). The radial component F_{rad} of the cutting force F_c is particularly responsible for the change of loading direction as it is acting in the opposite direction to the feed direction during the machining. In addition to cutting forces, the gravitational force and force resulting from the weight of a virtual spindle with the value of $F_v = 50$ N were included to the simulation model.

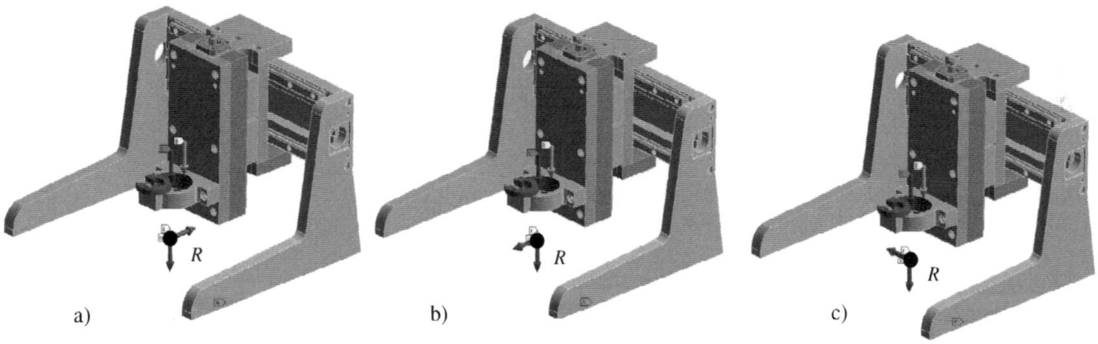

Figure 4. Schematic representation of acting loads in the direction of a) feed rate V_{f1}, b) feed rate V_{f2}, c) feed rate V_{f3}.

3. Results and discussion

The *Probe Deformation* was developed in the simulation model, to evaluate the effect of the structure deformation on the accuracy of the machining process under the specified load.This tool enables to measure the influence of the structure deformation on the displacement of a point created in the location of acting cutting force or in other words in the contact point of the milling tool and the workpiece. This point will be referred to as the point *R* (Figure 4).

The resulting deformations of the designed frame of construction caused by the machining with the feed rates of of V_{f1}, V_{f2} and V_{f3}.are illustrated in Figure 5, Figure 6 and Figure 7, respectively. The obtained results revealed a significant absence of stiffness, especially in the case of feed rate V_{f1} (Table 4).

Table 4. Computed displacements of due to the machining feed rate of V_{f1}, V_{f2} and V_{f3}.

Result	Feed rate		
	V_{f1}	V_{f2}	V_{f3}
Displacement of a point *R* in *X*-direction [mm]	−0.170	0.070	−0.048
Displacement of a point *R* in *Y*-direction [mm]	−0.150	0.034	−0.056
Displacement of a point *R* in *Z*-direction [mm]	-3.17×10^{-4}	2.67×10^{-4}	−0.060
Magnitude of the displacement of a point *R* [mm]	0.228	7.76×10^{-4}	0.095
Resulting deformation of the frame [mm]	0.230	0.075	0.103

IOP Conf. Series: Materials Science and Engineering **266** (2017) 012015 doi:10.1088/1757-899X/266/1/012015

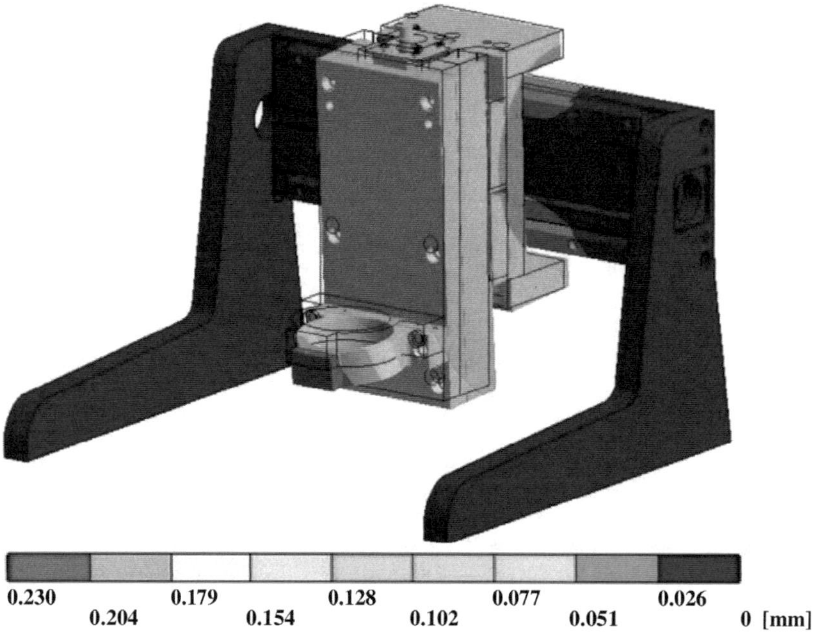

Figure 5. Deformation of the designed frame of construction caused by the machining with the feed rate V_{f1}

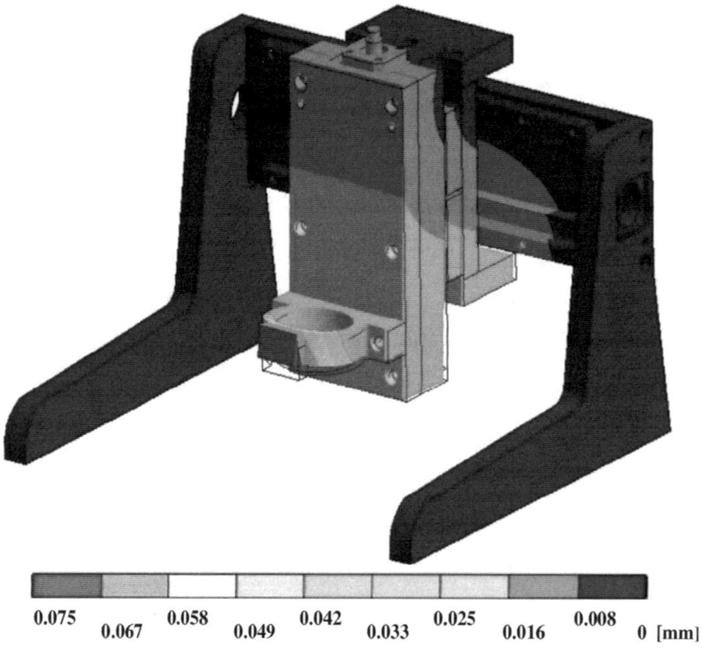

Figure 6. Deformation of the designed frame of construction caused by the machining with the feed rate V_{f2}.

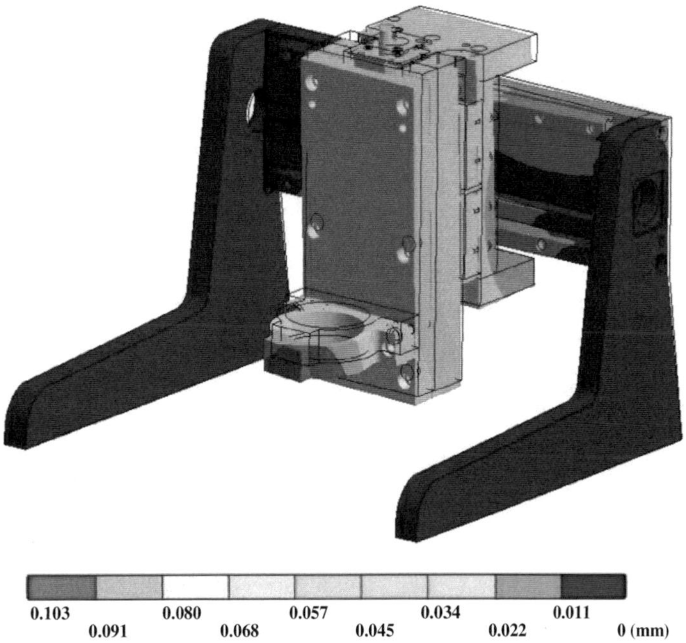

Figure 7. Deformation of the designed frame of construction caused by the machining with the feed rate V_{f3}.

The maximum deformation of the analyzed frame of construction was computed by the machining with the feed rate of V_{f1} at the level of 0.230 mm, while the displacement of the point *R* from its defined (ideal) position was 0.228 mm (Table 4). The deformations of the frame of construction resulting from the V_{f2} feed direction was the lowest. The maximum frame deformation in this case is 0.075 mm which is only 32% of the maximum value of deformation computed for the feed rate V_{f1}. The maximum deformation of the construction frame in the case of the feed rate V_{f3} was 0.103 mm representing 45% of the maximum deformation for the feed rate V_{f1}.

The highest value of von Misses stress at the level of 29.56 MPa was localized in the linear guides (*Y*-axis) and the spindle holder of the milling machine (Figures 8). The spindle holder is an important part of the machine tool construction. Results of numerical simulations indicate that the maximum stresses are concentrated right in this part. In this reason, the particular attention must be paid to this component when the design modification and changes will be accomplished.

Generally, proper changes and modifications of the geometry and/or materials applied for critical, the most loaded parts of milling machine would bring significant improvement of CNC machine stiffness. Aluminium alloys are used in the hobby segment in the largest range because of their good machinability and the low weight compared to steel (almost three times lower weight). The weight of the components is also determining for other properties and parameters of the machine. Moreover, the driving units with the less power can be used leading to the further reduction of the weight of the machine and its equipment. The optimal solution could be a hybrid construction consisting of combination of an aluminium alloy and steels that would be used for components subjected to higher loading. The modification of the supporting components can also be achieved by changing of the design using the ribbed components. Today, such design elements are often used in design of milling machines because the components with ribbed geometry generally better respond to static and dynamic loading [7].

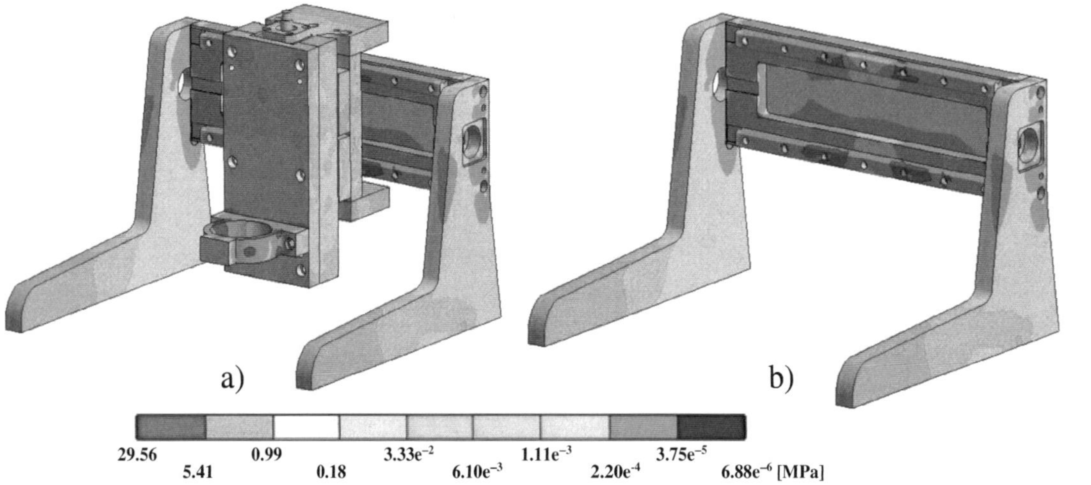

Figure 8. Distribution of von Misses stresses a) in the designed frame and b) in the linear guides due to the machining with the feed rate V_{f1}.

Based on the results of numerical simulations, not only the optimal shape of single components of a milling machine for the conditions in which the resulting product will be exploited can be designed but also the optimal geometric and shape arrangement of the functional elements of the product can be improved considering one or more selected parameters. The selected parameters can be for example the maximum performance, efficiency or the minimum price, weight, stresses, deformations, increase in service life or others. In the performed numerical simulation, the selected parameter was the deformation of the frame which further determined the machine accuracy.

4. Conclusions

The paper deals with the modelling and numerical simulation of loads arising in the milling process and their impact on the deformations and stresses developed in the frame of construction of hobby milling machine. For the static stress-strain analysis of the machine frame subjected to the different types of loading during a milling process, the simulation model was developed and used for the evaluation of deformations and stresses during the machining with the feed rates in different directions. In addition to the deformation of the machine frame, the accuracy of the milling tool positioning represented by the displacement of the reference point R was investigated.

As it follows from the results of numerical simulations, the maximum deformation of the machine frame at the level of 0.230 mm was attained for the machining with the feed rate V_{f1} in the direction of X-axis. The highest value of von Misses stress of 29.56 MPa was localized in the linear guides and the spindle holder of the milling machine. The displacement of the tip of milling tool in the reference point R from its defined position was 0.228 mm. It must be noted that in the direction of X-axis the machine has the maximum feed axis travel of 500 mm. The positioning of larger workpieces is assumed to be accomplished in such a way that the machine feed rate in this axis is exploited the most. From this point of view, it is necessary to revise the design of the machine with the particular attention on the machine stiffness in the direction of X-axis.

Acknowledgments

Research has been supported by the Programs to support excellent young researchers and research teams (EBW-AHSS 1387 and 1347/PRO-MAT-ZVAR) of STU Bratislava.

IOP Conf. Series: Materials Science and Engineering **266** (2017) 012015 doi:10.1088/1757-899X/266/1/012015

References

[1] Gajdoš J 2015 *The Future of CAD Systems* CAD **5**
[2] Marijn P, Zwier W and Wits W 2017 Physics in Design: Real-time Numerical Simulation Integrated into the CAD Environment *Procedia CIRP* **60** 98-103
[3] Autodesk Inc. San Rafael California USA
[4] SolidWorks SIMULATION Dassault Systèmes Vélizy-Villacoublay France
[5] Creo Simulate PTC - Parametric Technology Corporation, Needham, Massachusetts, USA.
[6] CATIA Dassault Systèmes Vélizy-Villacoublay France
[7] Hloužek J 2015 *ANSYS AIM- Physically linked analyzes in one program environment CAD* **5**
[8] Marek J 2014 *Construction of CNC machine tools 3* (Prague: MM publishing)
[9] Hong C C 2016 Static structural analysis of great five-axis turning–milling complex CNC machine *Engineering Science and Technology* **19** 1971-1984
[10] Liangchao Z, Ming L and Ralph R M 2016 *Direct simulation for CAD models undergoing parametric modifications* Computer-Aided Design **78** 3-13
[11] Łukaszewicz K 2017 Use of CAD Software in the Process of Virtual Prototyping of Machinery Procedia Engineering Journal **182** 425- 433
[12] Möhring H, Brecher C and Abele E 2015 Materials in machine tool structure *CIRP Annals – Manufacturing technology* **64** 725-748
[13] Rejt M and Píška M 2006 *Teorie obrábění, tváření a nástroje* (BRNO: Akademické nakladatelství CERM) **1** 225
[14] ANSYS Inc. Computer software Pennsylvania, United States

IOP Conf. Series: Materials Science and Engineering **266** (2017) 012016 doi:10.1088/1757-899X/266/1/012016

Corrosion behaviour of electropolished AISI 316L austenitic biomaterial in physiological solution

V Zatkalíková[1], L Markovičová[1] and M Škorvanová[2]

[1] University of Žilina, Faculty of Mechanical Engineering, Department of Material Engineering, Univerzitná 8215/1, 01026 Žilina, Slovakia
[2] Comenius University in Bratislava, Jessenius Faculty of Medicine in Martin, Malá Hora 10701/4A, 036 01 Martin, Slovakia

viera.zatkalikova@fstroj.uniza.sk

Abstract. Due to suitable mechanical properties, satisfactory corrosion resistance and relatively low cost, austenitic stainless steels are important biomaterials for manufacture of implants and various medical instruments and devices. Their corrosion properties and biocompatibility are significantly affected by protective passive surface film quality, which depends on used mechanical and chemical surface treatment. This article deals with corrosion resistance of AISI 316L stainless steel, which is the most widely used Cr-Ni-Mo austenitic biomaterial. Corrosion behaviour of five various surfaces (original, electropolished, three surfaces with combined treatment finished by electropolishing) is evaluated on the bases of cyclic potentiodynamic polarization tests performed in physiological solution at the temperature of 37± 0.5 °C.

1. Introduction

Currently, austenitic stainless steels are one of three metallic biomaterials used in the manufacture of implants. Coronary stents, fracture fixation plates and screws, spinal implant devices, aneurysm clips, temporary fixation devices, dentistry and surgical instruments are the most common medical applications of austenitic biomaterials [1, 2]. Compared to titanium alloys and chromium-cobalt alloys, the advantages of stainless steels are suitable mechanical properties (strength, mechanical workability), low cost and acceptable corrosion resistance [2-4]. Due to the protective passive surface film, stainless steels are resistant to the uniform corrosion [5, 6], but possibility of local pitting or crevice corrosion have to be taken into account in an internal human body environment [1, 2]. The surface film affects an interaction between the living tissue and implant. Therefore the quality of surface (largely influenced by used mechanical and chemical surface treatment techniques) plays the major role in the biocompatibility and corrosion resistance of implants [1-3].

Polished smooth surfaces are strictly required especially for surgical and dentistry instruments, ophthalmic devices, coronary stents, fracture fixation plates and screws [7, 8]. According to studies of many authors, bright smooth surfaces have markedly higher corrosion resistance [9-11] and also provide increased resistance to microorganisms that have less chance of getting caught [8] than those with higher roughness. For the above mentioned biomedical applications, mechanical and electrochemical polishing of austenitic steels is commonly combined with pickling (acid cleaning), ultrasonic cleaning, chemical passivation [8, 12].

Content from this work may be used under the terms of the Creative Commons Attribution 3.0 licence. Any further distribution of this work must maintain attribution to the author(s) and the title of the work, journal citation and DOI.
Published under licence by IOP Publishing Ltd

There are various polishing processes commonly used to obtain a smooth material surface. Traditional mechanical polishing may result in deformed layer and residual stresses on the treated surface [8]. Electropolishing is an electrochemical surface finishing process to enhance high surface brightness (mirror finish) with very low surface roughness ($R_a < 2$ μm), without residual surface tensions and with improved corrosion resistance [12]. A particular layer from the metal's surface is removed and a high quality passive oxide film is created.

This article deals with corrosion resistance of AISI 316L stainless steel, which is the most widely used Cr-Ni-Mo austenitic biomaterial. Corrosion behavior of five various surfaces (original without additional mechanical and chemical treatment, electropolished, ground + electropolished, pickled + electropolished, ground + pickled + electropolished) is tested. Evaluation of the corrosion resistance is based on the results of electrochemical cyclic potentiodynamic polarization tests performed in physiological solution at the temperature of 37 ± 0.5 °C. After tests, the surfaces are checked by optical microscope observation.

2. Experimental material

The experimental material AISI 316L is Cr-Ni-Mo austenitic stainless steel with the chemical composition shown in table 1. The steel was purchased in sheet (1000 x 1500 mm) of 1.5 mm thickness. Its treatment (marked as IIB/IIIC) was based on cold rolling, homogenization annealing (1040 – 1100 °C) and pickling after slightly smoothing rolling (www.italinox.sk).

Table 1. Chemical composition of AISI 316L stainless steel (wt. %).

Cr	Ni	Mo	Mn	N	C	Si	P	S	Fe
16.79	10.14	2.03	0.82	0.05	0.02	0.031	0.03	0.001	balance

Microstructure of experimental material (Figure 1) is created by polyedric austenitic grains with numerous twins, which could be created by annealing or by rolling.

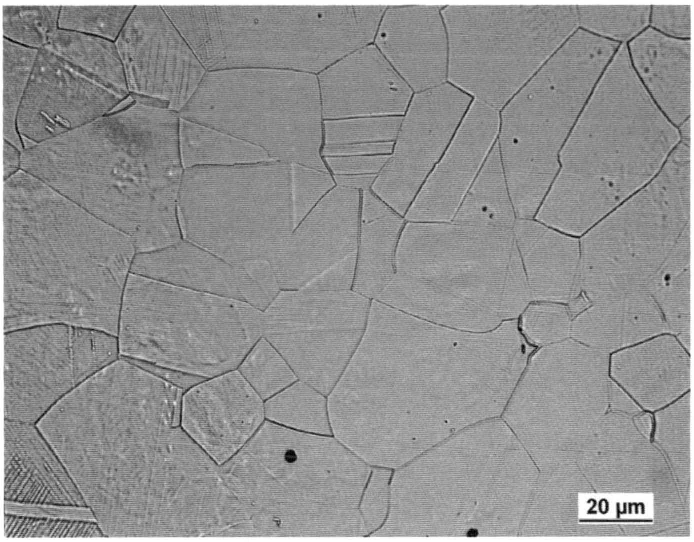

Figure 1. Microstructure of AISI 16L stainless steel, cross- section (Kallings etch.).

3. Experimental procedure

Five types of steel surfaces were used for potentiodynamic tests:

1. "as received" surface, i.e. original surface without additional mechanical and chemical treatment;
2. electropolished surface;
3. ground (1200 mesh) + electropolished surface;
4. pickled + electropolished surface;
5. ground (1200 mesh) + pickled + electropolished surface.

Electropolishing was performed in $H_3PO_4 + H_2SO_4 + H_2O$ electrolyte under the conditions listed in table 2 [13, 14]. The polished specimen was immersed into electrolyte and connected to the positive pole of the power source as an anode. A cathode was of the same material as the anode [13, 14].

Table 2. Conditions of electropolishing.

Component	Content (wt.%)	Temperature (°C)	Current density (A.cm^{-2})	Time (s)
H_3PO_4	64			
H_2SO_4	13	40	0.8	420
H_2O	23			

Pickling (acid cleaning) was performed under the conditions recommended for AISI 316L implants, namely coronary stents (Table 3) [7].

Table 3. Conditions of pickling (acid cleaning).

Component	Content (wt.%)	Volume (ml)	Temperature (°C)	Time (s)
HF	48	3		
HNO_3	65	9	22 ± 3	3600
H_2O	-	to 100 ml		

0.9 % sodium chloride solution was used as the corrosion environment for electrochemical cyclic potentiodynamic tests at the temperature of 37 ± 0.5 °C to simulate the internal environment of the human body.

Cyclic potentiodynamic polarization tests (ASTM G61) were carried out in a three electrode cell of corrosion measuring system (VoltaLab 10 with VSP unit). Polarization curves were obtained by the EC-LAB SOFT software. Potential between the sample and the electrolyte has been settled for 10 minutes before the polarization. Scan range was -0.3V – 1.2 V vs the open circuit potential and the scan rate was 1 mV/s. The surface of working electrode AISI 316L of 1 cm^2 area was rinsed with ethanol before measurement. The saturated calomel electrode (SCE) was applied as the reference electrode and platinum foil as a counter electrode. At least five experiment repeats were carried out for each specimen.

Surfaces of working electrode AISI 316L were observed by the optical microscope (Zeis Axio Imager A1m) after potentiodynamic measurements.

4. Experiment results and discussion

According to ASTM G61 standard method, the pitting potential (E_p) and the repassivation potential (E_{rp}) are the most important electrochemical characteristics of the cyclic potentiodynamic polarization curve. E_p potential can be determined as the potential of a sudden, strong increase of current density on the curve of the direct measurement, which denotes the breakdown of the passive film and the beginning of stable pit growth. The shift of E_p to more noble value means the rise of resistance to pitting [5, 6]. E_{rp} is obtained as the potential at which the backward measurement curve crosses the direct measurement curve.

If the passive film is highly resistant and current density is very law in whole passive region (according to Pourbaix diagram for Cr in neutral 0.9 % NaCl solution, the passive region reaches up to potentials 0.7 - 0.8 V) [15], it means that the pitting corrosion is not initiated during measurement. In this case the cyclic curve is not of typical loop shape and the current density increases at the trans-passivation potential (E_t). If the potential overruns E_t value, the surface passive film is dissolved, and an active corrosion of material starts (Cr^{3+} is converted to Cr^{6+}) [5, 15].

The corrosion potential (E_{corr}) and the corrosion current density (i_{corr}) are additional electrochemical characteristics which can be determined according to Tafel analysis of the potentiodynamic curve [15]. The lower E_{corr} points to the lower thermodynamic stability of material. The i_{corr} value has the relationship to the kinetics of the corrosion process. The higher i_{corr} value means the higher corrosion rate.

The cyclic potentiodynamic curves of tested AISI 316L surfaces are shown in Figures 2 – 4. The obtained electrochemical characteristics are presented in Table 4.

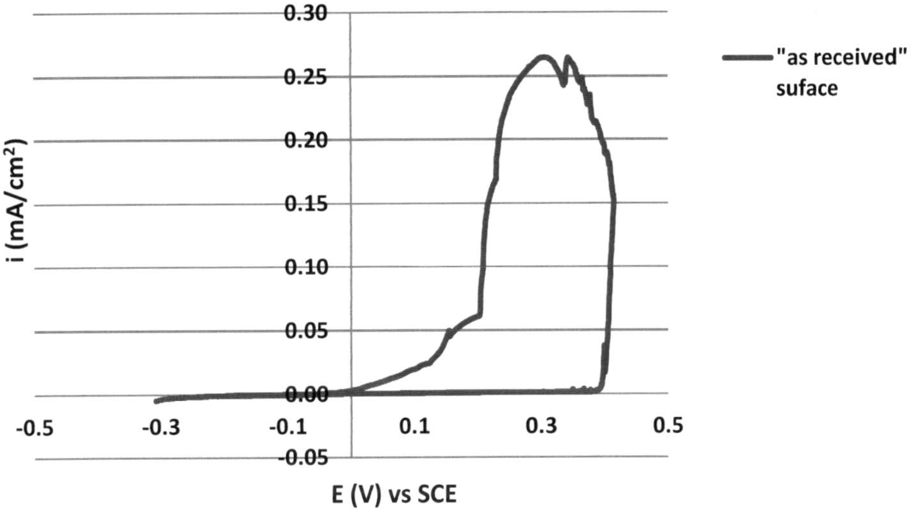

Figure 2. Cyclic potentiodynamic curve of AISI 316L working electrode with "as received" surface.

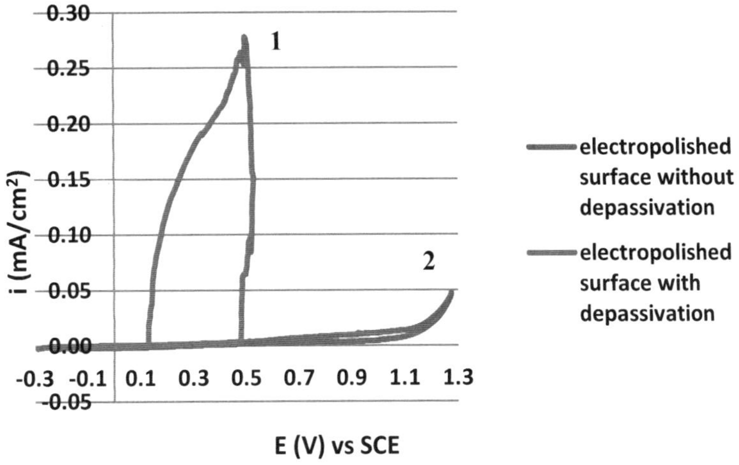

Figure 3. Cyclic potentiodynamic curves of AISI 316L working electrode with electropolished surface.

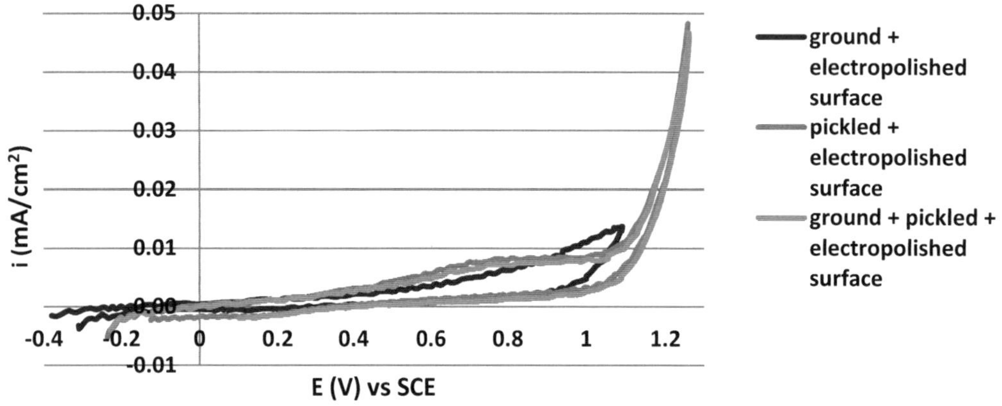

Figure 4. Cyclic potentiodynamic curves of AISI 316L working electrode with combined surface treatment finished by electropolishing

The "as received" surface curve (Figure 2) is of a typical loop shape with clearly observable pitting potential. According to electrochemical characteristics (namely E_p, table 4) this type of surface is least thermodynamically stable compared to other tested surfaces. The pitting corrosion damage was visible to the naked eye and also observed by an optical microscope (Figure 5A).

Table 4. The electrochemical characteristics of tested AISI 316L surfaces.

Type of surface		E_p (V)	E_{rp} (V)	E_t(V)	E_{corr} (V)	i_{corr} (μA/cm^2)
"as received"		0.381	−0.052	-	−0.071	0.078
electropolished	1	0.476	0.128	-	−0.069	0.062
	2	-	-	1.121	0.027	0.035
ground + electropolished		-	-	-	−0.108	0.044
pickled + electropolished		-	-	1.060	−0.035	0.265
ground + pickled + electropolished		-	-	1.103	0.059	0.016

The electropolishing of the original surface (Figure 3) improved the quality of the passive layer, but obtained curves fluctuated between the typical loop shape (1) and the shape with low current density, without depassivation in passive region and with an increase of current density at the trans-passivation potential E_t (2). The passive film was apparently not uniform enough, chloride ions could penetrate through the weakened places and the stable growing corrosion pits were formed. The course of the measured curve depended on the particular area selected for the potentiodynamic test. Similar curves measured for the elctropolished surface of the same steel were also found by the authors [14].

The typical corrosion pit on the electropolished surface is shown in Figure 5 B. The high corrosion resistance was reflected on surfaces that have been mechanically/chemically treated before electropolishing (Figure 4). The obtained curves were without depassivation in passive region and with transition to the transpassivity state. The display of the respective curves at one scale and their comparison showed some differences in the passive state current density and in approximate values of the E_t potentials.

It is worth recalling that the threshold value of the current density signaling the transition from the passivity state to the active dissolution of the metal is usually assumed to be 0.05 mA / cm^2 [6]. Generally, a passive state with high stability exhibits a low passive current density without large fluctuations on the direct measurement curve and the transition to the transmissivity (reflected by a significant increase in current density) at the high E_t potential [5].

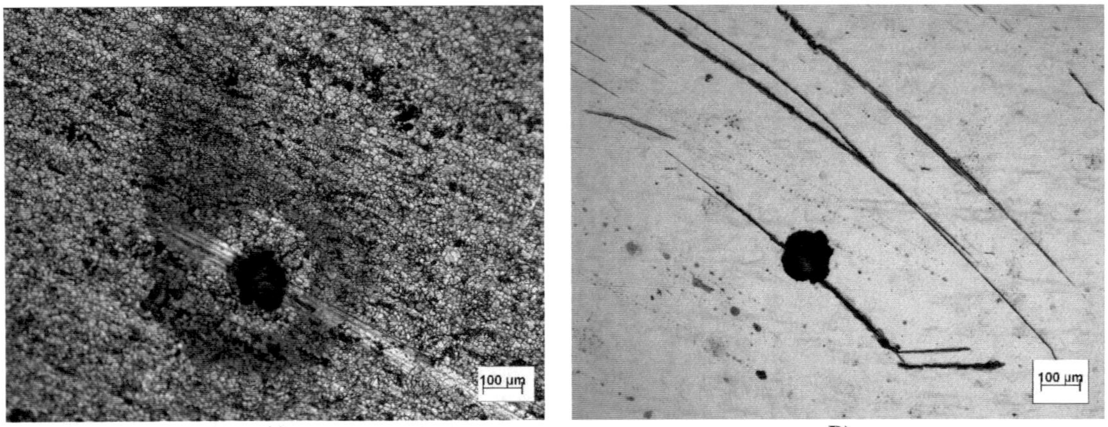

A) **B)**

Figure 5. Typical pit shape and size on the steel surface after cyclic potentiodynamic test: A) "as received" surface, B) electropolished surface.

Taking into account the above mentioned facts, the ground + pickled + electropolished surface exhibits the best resistance to pitting. On the contrary, grinding + electropolishing can be considered the worst from three surface treatment tested combinations. On the ground + electropolished surface the passive state appears the least stable - current density slightly increases throughout the passive area (Figure 4). Microscopic observation of these combine- treated surfaces after cyclic potentiodynamic tests did not reveal any pitting corrosion damage (Figure 6).

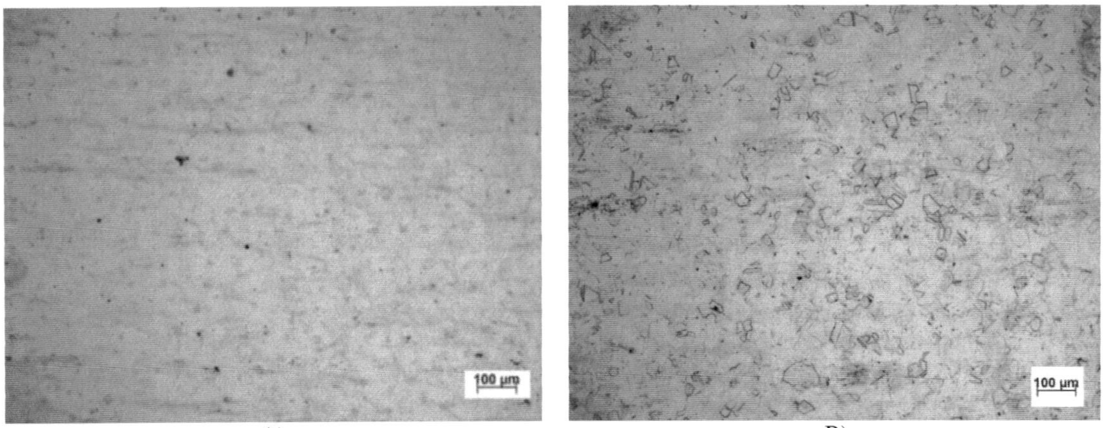

A) B)

C)

Figure 6. Typical steel surface after cyclic potentiodynamic test:
A) pickled + electropolished surface,
B) ground + electropolished surface,
C) ground + pickled + electropolished surface.

Very good quality of electropolished surfaces is probably connected with the newly grown, uniform and compact oxide film with higher chromium and oxygen content [16]. The authors [3] documented chromium enrichment in surface passive film of AISI 304 steel after electropolishing. According to the authors [3, 16], the increase of Cr/Fe ratio is caused by selective dissolution leading to the formation of a passive layer during the electropolishing process.

5. Conclusion

Based on the results of performed experiments can be concluded:

- The "as received" surface state is least thermodynamically stable. The pitting corrosion damage was visible to the naked eye.
- Single polishing without pre-treatment does not appear as a surface treatment that would provide a sufficiently uniform and resistant passive film.
- Surfaces with combined treatment finished by electropolishing (ground + electropolished, pickled + electropolished, ground + pickled + electropolished) exhibit better corrosion resistance than single electropolished surface. The least suitable combination is grinding + electropolishing. The best combination is grinding + pickling + electropolishing, which provides the highest corrosion resistance.

Potentiodynamic tests are convenient and fast, but at least two independent tests are needed to assess clearly the corrosion resistance of a given biomaterial (especially when intended for long-term implantation). Although the surface is properly treated, it is important to avoid mechanical damage during implantation (e.g. by surgical instrument). The passive film breakdown could cause dangerous local corrosion in the internal environment of the human body.

Acknowledgement

The research was supported partially by Scientific Grant Agency of Ministry of Education, Science and Sport of Slovak Republic and Slovak Academy of Sciences, grant VEGA No. 1/0683/15 and grant KEGA No. 049ŽU-4/2017.

References

[1] Ratner B D, Hoffmann A S, Schoen F J and Lemons J E 2004 *Biomaterials Science* Elsevier, San Diego USA
[2] Nagase M 1995 *Host Reactions to Particulate Biomaterials. Encyclopedic handbook of biomaterials and bioengineering* **1** Part A New York
[3] Ghanavati S, Shishesaz M R, Farzam M and Danaee I 2015 Effects of Surface Treatment on Corrosion Resistance of 304L and 316L Stainless Steel Implants in Hank's Solution *Iranian Journal of Oil & Gas Science and Technology* **5** 1 65-72
[4] Uhríčik M, Oravcová M, Palček P, Sapieta M and Chalupová M 2016 *The stress detection and the fatigue lifetime of stainless steel AISI 316L during three-point bending cyclic loading* (Conference Paper) EAN 2016 - 54th International Conference on Experimental Stress Analysis
[5] Szklarska–Smialowska Z 2005 *Pitting and crevice corrosion* NACE International Houston Texas
[6] Liptáková T 2009 *Bodová korózia nehrdzavejúcich ocelí. (Pitting corrosion of stainless steels.)* EDIS Žilina
[7] Zhao H, Van Humbeeck J, Sohier J and De Scheerder I 2002 Electrochemical Polishing of 316L Stainless Steel slotted tube coronary stents *Journal of Materials science: materials in medicine* **13** 911-916
[8] Lee S J and Lai J J 2003 The effects of electropolishing (EP) proces parameters on corrosion resistence of 316L stainless steel *Journal of Material Processing Technology* **140** 206-210.

[9] Liptáková T, Alaskari A and Halamová M 2012 Surface treatment of the AISI 316L after welding and study of corrosion behaviour *Periodica Polytechnica Transportation Engineering,* **41** 2 143-147

[10] Hong T and Nagumo M 1997 Effect of surface roughness on early stages of pitting corrosion of Type 301 stainless steel *Corros. Sci.* **39** 9 1665–1672

[11] Sasaki K and Burstein G T 1996 The generation of surface roughness during slurry erosion-corrosion and its effect on the pitting potential *Corros. Sci.* **38** 12 2111–2120

[12] Núñez P J, García-Plaza E, Hernando M and Trujillo R 2013 Characterization of Surface Finish of Electropolished Stainless steel AISI 316L with Varying Electrolyte Concentrations *Procedia Engineering* **63** 771-778

[13] Vaňa D 2014 *Porovnanie technologických možností plazmového a elektrochemického leštenia antikoróznych ocelí v elektrolyte* AlumniPress Trnava

[14] Oravcová M, Palček P, Zatkalíková V, Tański T and Król M 2017 Surface treatment and corrosion behaviour of austenitic stainless steel biomaterial *IOP Conference Series: Materials Science and Engineering* **175** 1

[15] Hadzima B and Liptáková T 2008 *Základy elektrochemickej korózie kovov (The bases of electrochemical corrosion of metals)* EDIS Žilina

[16] Hashimoto K, Asami K, Kawashima A, Habazaki H and Akiyama E 2007 The role of Corrosion-resistant alloying Elements in Passivity *Corros. Sci.* **49** 1 42-52